DISCUSSION OF THE METHOD

Conducting the Engineer's Approach to Problem Solving

Billy Vaughn Koen

New York   Oxford
OXFORD UNIVERSITY PRESS
2003

Oxford University Press

Oxford   New York
Auckland   Bangkok   Buenos Aires   Cape Town
Chennai   Dar es Salaam   Delhi   Hong Kong   Istanbul   Karachi
Kolkata   Kuala Lumpur   Madrid   Melbourne   Mexico City   Mumbai   Nairobi
São Paulo   Shanghai   Taipei   Tokyo   Toronto

Published by Oxford University Press, Inc.
198 Madison Avenue, New York, New York, 10016
http://www.oup-usa.org

Oxford is a registered trademark of Oxford University Press

Library of Congress Cataloging-in-Publication Data
Koen, B. V.
  Discussion of the method : conducting the engineer's approach to problem solving
/Billy Vaughn Koen.
    p. cm.
  Includes index.
  ISBN 0-19-515599-8.
    1. Methods engineering. 2. Problem solving. I. Title.

T60.6 .K635 2003
620—dc21

                                                                            2002193018

Printing number: 9 8 7 6 5 4 3 2 1

Printed in the United States of America
on acid-free paper

For Deanne, Kent, and Doug

CONTENTS

FIGURES AND TABLE

FIGURES

TABLE

COLOR INSERT

Found between pages 148 and 149.

PREFACE

This book is about the universal method. It is about what you do, what I do, what every human has done at every minute of every day since the birth of humanity. Would you change any aspect of your world? If so, you need a universal method to guide you along the steps you must take.

Philosophers have long sought a universal method, mostly to no avail. They failed to look for it in the most unlikely of places, in the engineering method. If there is a universal method—and there is—at a minimum it should be, well, universal. Whether we look to philosophy, to science, to the arts, to engineering, or elsewhere, any method that claims to be universal must subsume all of its competitors. Perhaps surprisingly, the engineering method will emerge as the prototype in our search for the universal method. This choice is defensible because the engineering method is undeniably an effective method. We validate it every time we drive a car, pick up the telephone, sit down at a computer, cross a bridge, take an airplane, turn on the lights, take our medicine, paint our house, put on a sweater, or read a book. If a universal method is to exist, certainly the engineering method is a proven, practical, and important exemplar. The engineering method captures all of the characteristics of a universal method that you will need to create the world you desire. The objective of this book is to define the engineering method and to generalize it to the universal method.

Although there is one universal method, there are many implementations of it. You are absolutely unique—whether by hopes, dreams, areas of interest, profession, race, customs, religion, or nationality. It is not surprising, then, that each person who reads this book will read a different one and would read still another one if the book were reread. Characterizing the uniqueness of each of us is an important and essential part of this book. The universal method defines our separate humanities.

The universal method is the result of a hunt that began in the spring of 1965. Now thirty-six years later, my conviction of having found a universal method is unwavering, if not stronger. All that remains is for you to join me in examining this once elusive prey.

Editor's Note: The author has modeled this book on Descartes' classic *Discours de la Méthode*. The reader will detect echoes of the style, words, and structure of the original, as well as echoes of the works of other well-known scholars with several important exceptions. First, although the original contained no references, the present volume contains only sufficient information to allow an expert in the field to locate easily the cited material while not delaying those who are not specialists. Second, at times the unconventional typography of underlining certain letters appears to remind the reader of the unique characteristics of certain words. An understanding of the theoretical necessity for this unusual rhetorical device will become apparent midway through the book. Both of these departures from modern practice reinforce the central theme of this book.—Ed.

INTRODUCTION

You and I are participating in a magnificent experiment to see whether Nature's latest wrinkle, the human species armed with its new weapon, intelligence, has survival value. Considering the 100 million species of animals that are estimated to have become extinct since life first appeared on this planet and the large number of other species such as the opossum and cockroach that have survived essentially unchanged for millions of years, even a preliminary celebration by *Homo sapiens*, with its paltry three hundred thousand years of existence, is premature. If this experiment fails (and the mental health of our species, unbridled terrorism, genetic engineering, pollution, and overpopulation suggest it may), I doubt that Nature will give more than a majestic shrug.

This disinterest is not good enough for me, and I doubt it is good enough for you. As an engineer, I am interested in the most effective way to cause the changes I feel are desirable. Would you establish global peace or squelch an impending family feud? Would you improve your nation or modify your local government? Would you feed the one quarter of the world population that goes to bed hungry or decide what to serve for dinner? If so, the world is not as you would have it, and you too are interested in a method for change.

The overall goal of this book is to develop a compelling definition of the *universal method*—that is, to define the method you should use, indeed, the one you must use, to solve the problems you would solve. During our time together we will first discover that the single most incisive strategy for causing change ever devised is the engineering method. Then the engineering method will be used as a model for the universal method. This generalization to the universal method is the strategy to which you should turn to create your private utopia.

While the study of the engineering method is important *to create* the world we would have, its study is equally important *to understand* the world we do have. Our environment is a collage of engineering problem solutions. Political alliances and economic structures have changed dramatically as a result of the telephone, the computer, the atomic bomb, and space exploration—all undeniably products of the engineering method. Look around the room in which you are now sitting. What do you find that was not developed, produced, or delivered by the engineer? What could be more important than to understand the strategy for change whose results surround us now and, some think, threaten to suffocate, to pollute, and to bomb us out of existence?

Although you speak freely of technology, it is unlikely that you have the vaguest philosophical notion of what it is or what is befalling you as it soaks deeper into your life; and, if you are typical, you are becoming apprehensive. What if I were to ask you, "What is the *scientific* method?" You would undoubtedly be able to answer. You might respond, "Science is theory corrected by experiment" or the other way around. If you are a specialist in the scientific method, you might refer to the theory of falsification or the theory of paradigm shifts. But when asked, you, or anyone else for that matter, whether layperson, scientist, or specialist in the history of science, would feel qualified to give a cogent response. Now, as you sit immersed in the products of the engineer's labor and are becoming concerned with them, I ask you: "What is the *engineering* method?"

Your lack of a ready answer is not surprising for a variety of reasons. Unlike the extensive analysis of the scientific method, little significant research to date has sought the philosophical foundations of engineering. Library shelves groan under the weight of books by the most scholarly, most respected men that analyze the human activity called science. Among many, many others we find the work of the Ionian philosophers (Thales, Anaximander, and Anaximenes), in which many feel the germ of the scientific method was first planted in the sixth century B.C.; of Aristotle in the *Organum*; of Bacon in the *Novum Organum*; of Descartes in the *Discours de la Méthode*; of Popper in *The Logic of Scientific Discovery*; and of Kuhn in the *Structure of Scientific Revolutions*. No equivalent body of research treats the engineering method. No equivalent philosophy of engineering exists.

To identify a second reason for the lack of understanding of the engineering method, consider the professions that affect your daily life, such as law, economics, medicine, politics, religion, and science. For each you can easily name at least one person who is well known to the general public as a wise, well-read scholar in the field—a person to whom you could turn to put her profession in perspective. Now name an engineering scholar with similar qualifications. No profession affecting the world to the extent of engineering can claim this isolation.

As a general rule, no argument is improved by exaggeration because the listener immediately becomes suspicious of what is to come. Therefore, my claim that we lack scholarly spokesmen for engineering must be properly interpreted and not taken as overstatement. The present point is not, as some have held, that engineers are never scholarly nor well read. This argument cannot be sustained because many engineers have produced scholarly works. For instance, Ludwig Wittgenstein, a famous philosopher whose principal concern was the relationship between language, mind, and reality, formerly studied mechanical engineering in Berlin and worked in aeronautical research at the University of Manchester. Even as a philosopher, Wittgenstein's *language-games, nest of propositions,* and *world-pictures*, along with his insistence on the importance of context, reveal his training as an engineer. Another mechanical engineer, Alexander Calder, achieved world renown for introducing motion into the world of art with his invention of the mobile. Among my colleagues, one, originally an engineer, is now a recognized expert on Faulkner, and another deserted engineering for Emerson. In addition, two former civil engineers have won the Nobel Prize for Literature: Salvatore Quasimodo, an Italian, for poetry, and José Echegaray, a Spaniard, for drama. Of neither can we argue that engineering was a minor, unimportant activity in his life. Each rose to prominence as an engineer before turning to the humanities. Two Nobel Prizes for Literature out of the limited number that have been awarded is certainly impressive and destroys the notion that engineering and scholarly abilities are necessarily mutually exclusive. To be precise then, my challenge to you is to name an engineer who is wise, well known, well read, and scholarly in his role as an engineer. That is to say, in the event of a serious nuclear incident, the failure of the pylons on a large airplane, the pollution of ground water by chemical wastes, or the failure of a walkway in a modern hotel, to whom could you or the news media turn to put the situation in perspective and settle your fears? The inevitable answer: There is no one.

Unfortunately the situation is far worse than just the lack of an engineering spokesman. Remembering that (1) few high school students take courses in engineering; (2) the study of technology is not required for a liberal arts degree; and (3) sociologists, psychologists, historians, and religious fanatics, not engineers, write most of the pro- and anti-technology literature—are you sure, as an engineering spokesman speaks of optimization, factors of safety, and feedback, that you would understand him even if he were to exist? With little research into the theory of engineering, with no scholarly, well-known spokesman for engineering, and with the exclusion of engineering from the general education requirements of the public, coupled with the chronic aversion of engineers to write about their world, the existence of people who do not understand the engineering method and are a bit frightened by technology is not really too surprising.

Classification is a risky business, and classification of a method as being *the* engineering method even more so. Nature is adroit in creating objects whose only excuse for being seems to be to thwart someone's effort to put them into a category. Even such self-evident categories as alive or dead, animal or plant, and male or female seem to defy analysis. Is the DNA molecule alive or dead? Are Euglena (one-celled animals with chloroplasts) or *Pfiesteria piscida* (animals that masquerade as plants) plants or animals? Are XXY individuals, striped sand bass (which have male and female sex organs), clown fish (which spontaneously turn from male to female), or sexually dimorphic animals male or female? Even the apparently straightforward task of classifying the writings of another as philosophy, autobiography, imaginative literature, fiction, nonfiction, or roman à clef at times proves unexpectedly difficult. Since the introduction of *category* as a technical word by Aristotle, the battle has raged as to what one is and how many there are. I do not expect, therefore, to be able to define the engineering method and differentiate it from the methods of common sense, of science, and of philosophy in a manner that will please all. The boundaries between each of these concepts are just too fuzzy. The specific difficulty of trying to define a method as the engineering method is not of crucial concern, however, because as promised earlier, ultimately this discussion will assert that only one universal method exists—a method that subsumes all methods, including all of those mentioned here, in a new liberal synthesis. The initial emphasis is on the engineering method only because it represents the most conspicuous and accessible model of this universal method.

Thank you for the compliment of choosing this book, carrying it home under your arm, and inviting me into your home. By you I mean you, the real person sitting opposite me holding this book in her hand. As I understand the rules of thumb for good writing, the author should never embrace his audience too closely. The reader, it seems, is more at ease standing somewhat apart and resents being given a bear hug. But I have no taste for creating a proscenium between us: teasing you with a false sense of distance; tormenting you with impersonal, convoluted arguments that require more effort to unscramble than they are worth. That is not the style of the engineer. I prefer to believe that you and

I are sitting together, casually dressed, after finishing a good meal (Scandinavian, I think) and have settled down for a few minutes' conversation. At times our mutual friends, René, Kurt, Sextus, Immanuel, among others, will drop by to join us and you will have the pleasure of meeting some of my more contemporary mentors, Johnny, Tommy, Kent.

It was always like this at my grandmother's house in the Southwest. Neighbors were welcome to "sit a spell" and enjoy a friendly fire. The conversation was punctuated by long silences accompanied by a great deal of rocking, each of us with his own thoughts. No one considered himself an expert; no one felt secure enough in his own thoughts to deliver a definitive discourse; no one was concerned about a formal treatise on anything. Everyone was only interested in an informative discussion. Mulling over a new idea, perhaps to ultimately reject it, was the most pleasurable part of the evening. As you and I think together, please put this book down occasionally—and rock.

My major concern is that I do not know your name or what you do for a living. Philosophers, novelists, and engineers observe the world from different seats in the balcony, and, if they do not recognize what they see through their own opera glasses, give the performance a bad review irrespective of its intrinsic value. The same situation is true for a book. The philosopher will reject one that does not contain an occasional *ad rem*; the novelist, one without an ablative absolute or two. The equivalent engineering view was expressed by one of my engineering professors who, quite seriously it turns out, held that no book should ever be printed that did not contain at least one mathematical equation. This engineer's view is so odd that I am glad that my wife was present to validate the exchange. There is little wonder that some consider the engineer to be pathologically illiterate. But what should we make of the parallel narrowness of the philosopher and novelist who cannot solve the simplest calculus problem? Or, the report by Robert Frost in *Education by Poetry, A Meditative Monologue*, "I once heard of a minister who turned his daughter—his poetry-writing daughter—out on the street to earn a living, because he said there should be no more books written; God wrote one book, and that was enough." My desire is to establish a philosophical basis for the engineering method comparable to that of the scientific method, to generalize it to a universal method, and to suggest its application to the problems we confront—that is, to position engineering along the spectrum of human knowledge. But since we seek a universal method, this discussion must be readable by all, whether philosopher, novelist, or engineer.

One of my distant relatives, I think by marriage, who is a highly celebrated and gifted American author held that the greatest merit of style is to make the words disappear into thought. This must only be taken as a rule of thumb, however, because, although it is plausible and based on experience, it is potentially fallible. In fact, in the present discussion, style is of the essence. Make no mistake—this book was especially crafted so that each reader would read a completely different book—and read still a second one the second time around. Subtle changes in the rhetorical profile presage the opening and closing of different worlds. To the philosopher and engineer I whisper my concern: I hope you will withhold censure of this discussion till you have read it through. I am concerned that you might lay it aside due to a mistake I have made in its design in relation to your own particular background.

As for myself, I have never supposed that I were a better engineer than the average. On the contrary, I have often wished that my interests lay in one small, highly technical area instead of ranging over the strengths and weaknesses of the engineering method, its importance, and its interrelationship to the sciences and humanities. Neither do I claim that I am original in any of these thoughts, but only that I do not hold them because a personal, preconceived notion of absolute reason persuades me of their truth. Rather, I put them forth because they represent, for me, only a small compelling change in the thinking of some of the best minds in engineering.

I have had the opportunity to attend one of the most celebrated engineering schools in the world, where there would certainly be wise engineers if there were wise engineers anywhere on earth. The Japanese use the term *borrowed scenery* for the technique by which the size of a small garden is made to appear larger by so designing it that neighboring trees and shrubs are effectively absorbed into the local landscape. Better that I present these few thoughts on method as an autobiography of an idea, enhanced by the borrowed scenery of my former professors, of colleagues, and of the scholars I have had the good fortune to have read, since everything of importance found here can be traced back to a precise statement made by one of these great men. If you think you hear the echo of a phrase from the past, you are undoubtedly correct. Hopefully I shall weave their greatness into your world as it has been woven into mine. Even this discussion would have never taken place without the help and encouragement of Drs. Koerner, Hoberock, Upthegrove, Gordon, Smith, Wulf, Truxal, and Sloan, who read a draft from beginning to end.

If our discussion seems too long to be read in one sitting, it may be divided into six parts. The first three are concerned with the engineering method and are important in establishing a sound basis for this approach to problem solution; the last three are concerned with the universal method, using the engineering method developed in the first three parts as a model. Specifically, the objectives of the various parts of this discussion are as follows.

- **Part I**   **Some Thoughts on Engineering:** This part describes the problem situation that calls for the talents of the engineer and emphasizes how frequently the human encounters this kind of situation.

- **Part II**   **The Principal Rule of the Engineering Method:** Here the engineering method is defined.

- **Part III**   **Some Heuristics Used by the Engineering Method:** This part reexamines the important definitions we have developed and lists examples of the techniques the engineer uses to implement his method. It also describes several alternative definitions of the engineering method and puts our definition in its final, preferred form.

- **Part IV**   **The *Universale Organum*:** This part generalizes the engineering method to the universal method.

- **Part V**   **Summary of the Method:** At last a concise, philosophically justifiable statement of the universal method is possible.

- **Part VI**   **Application of the Method:** We conclude our time together with a specific example of the use of the universal method and my reasons for writing this book.

Some Thoughts on Engineering

T he use of the engineering method rather than the use of reason is humankind's most equitably divided endowment. I mean by

the *engineering method* the strategy for causing the best change in a poorly understood situation within the available resources

and by *reason*, the ability to distinguish between the true and the false, or what
Descartes has called "good sense." Whereas reason had to await early Greek
philosophy for its development, is even now denied in some cultures, and is in
retreat in others, the underlying strategy that defines the engineering method
has not changed since the birth of humans. *To be human is to be an engineer.*

The objective of this chapter is to prepare the way for a consideration of the
strategy the engineer uses to solve problems that will be given in Part II. To
achieve this objective we will divide our thoughts into three sections. In the first,
we will view the engineer through the eyes of the layperson, historian, etymol-
ogist, lexicographer, and anthropologist to learn how we are to recognize an engineer when we meet one. In the second, attention shifts to the characteristics
of a problem that requires the talents of this new acquaintance. Finally, in the
third part of this chapter, we will examine several problems that unexpectedly
bear a resemblance to engineering problems. We should anticipate that they
might be effectively attacked using the engineering method. Incidentally we will
come to appreciate how long ago it was that ancient man* first encountered engineering problem situations and how frequently modern man confronts them
today. Acceptance of the claim of identity between human and engineer will
come subliminally as the engineering method becomes more fully understood.
Consideration of the correct epistemological status of reason, the engineering

---

*The rules of contemporary writing in America require the author to go through
wild contortions to avoid the sexist use of masculine nouns and pronouns. This
was not the state of the art in 1982 when the first version was completed. With an
apology to those on both sides of this question, we will retain the original format
that referred to the engineer as male. At least initially, the pronoun *she* will refer
to everyone else: philosophers, scientists, artists, poets, and so forth. This serves as
an excellent example of the changing rules of thumb in acceptable practice over
time.

method's chief rival as a basis of the universal method, will be reserved until a much later chapter in this discussion.

THE ENGINEER

Most people think of the engineer in terms of his artifacts instead of his art. As a result they see diversity where they should see unity and find it hard to accept the identity of man and engineer. The question, "What is an engineer?" is usually answered by "a person who makes chemicals, airplanes, bridges, or roads." From the chemicals, the layperson infers the chemical engineer; from the airplanes, the aeronautical engineer; and from the bridges and roads, the civil engineer. Not only the layperson, but also the engineer, makes this mistake. Because the connection of the engineer with his completed design is so enduring and the connection with his use of method so fleeting, a person insists he is an engineer based on what he produces, irrespective of how he goes about it, instead of insisting that he is an engineer based on how he goes about it, irrespective of what he produces. In a similar fashion, the historian uses the existence of dams on the Nile, irrigation canals in various parts of the ancient world, gunpowder, and pottery to infer the existence of engineers and craftspersons in past civilizations. But behind each chemical, each road, each pot hides the common activity that brought it into being. It is to this unity of method that we must look to see the engineer in every man.

To be sure, we occasionally see an emphasis on method instead of object, ironically most frequently in the writings of the nonengineer. We read in the Sunday newspaper supplement that a reigning prince *engineered* the divorce of his daughter, in the daily newspaper that the clergy in Iran *engineered* the firing of the president, in a medical report that doctors have *engineered* a better bacterial host, and in a book on the game of chess that white has *engineered* a perfect counter to black's opening. Even the headline of one of Dear Abby's newspaper columns offers the advice, "Don't *engineer* a problem; just keep silent about past." Similar statements using the word *engineering* in the sense of creating a desirable change in an uncertain situation within the available resources are found daily in novels, reviews, and newspapers and heard on radio and television.

In each of these cases we sense that the word is being correctly used, and we are right. According to one of England's most noted nineteenth-century engineers, Sir William Fairbairn, quoted in *Technology and Change* (ed. by Burke and Eakin),

> The term engineer comes more directly from an old French word in the form of a verb—*s'ingénier* . . . and thus we arrive at the interesting and certainly little known fact, that an engineer is . . . anyone who seeks in his mind, who sets his mental powers in action, in order to discover or devise some means of succeeding in a difficult task he may have to perform.

The dictionary concurs by authorizing the verb *to engineer* as "to contrive or plan usually with more or less subtle skill or craft" and by giving as an example "to engineer a daring jailbreak." It also gives a second definition: "to guide, manage, or supervise during production or development," including the example "engineering a bill through congress." But despite these frequent examples to the contrary, when a definition of engineering is sought, the usual tendency is to look to concrete objects such as chemicals, airplanes, bridges, or roads instead of to the method that brought each of these engineering devices into existence.

This same confusion between art and artifact exists in efforts to date the birth of humankind. Most anthropologists define the human by his use of tools, none as eloquently as Loren Eiseley in his book *The Firmament of Time:*

> Massive flint-hardened hands had shaped a sepulcher and placed flat stones to guard the dead man's head. A haunch of meat had been left to aid the dead man's journey. Worked flints, a little treasure of the human dawn had been poured lovingly into the grave. And down the untold centuries the message had come without words: "We too were human, we too suffered, we too believed that the grave is not the end. We too, whose faces affright you now, knew human agony and human love."

In this miniature drama we see a problem born of the most human craving to create an appropriate resting place for a loved one and a solution pitifully constrained by severe limits in knowledge and resources. Recalling the definition of engineering method as the strategy for causing the best change in a poorly understood situation within the available resources, can we doubt that at least as early as Neanderthal man we find engineers?

In dating the birth of humankind, other scholars argue that it is more important that "primitive man stuck feathers in his hair" than that he worked in flint. We are told that our error in defining humans by their use of tools results from society's present preoccupation with materialism and technology on the one hand and the lack of traces of early myths, customs, and literature in the physical record on the other. "If only we could still see the pictures of earliest man and hear him sing," the humanist laments, "we would define him by his arts instead of his tools." Again the emphasis is on an object such as a glyph or a song, but in this case it is just a more ephemeral one.

Each of these views is too narrow and self-serving to be a proper definition of humankind. For behind the earliest crude flint, behind the earliest artistic scrawl lies a common process that brought it into being. We do not see this method because its earliest use, like the earliest picture or song, left no traces. But a common sense of method—of desirable change in an unknown, resource-limited world—pervades and certainly predates both the tools we find and the arts we infer. Throughout history, differences may have been observed in how well humans have used the engineering method or in the subject matter they have treated, but the use of the engineering method is coterminous with any reasonable definition of the human species. *The engineering method, like a midwife, was present at the birth of humankind.*

CHARACTERISTICS OF AN ENGINEERING PROBLEM

The engineer's method itself, not the results of this method, legitimates the word *engineer* in all of the previous examples, and it is by this method that the engi-

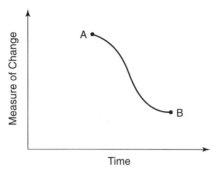

Figure 1 Measure of Change

neer should be recognized. Whether we consider humans at their birth, Neanderthals, the ancient dam builders, the prince who wants the divorce of his daughter, or the designer of a modern bridge, each is correctly called an engineer because each finds himself in an equivalent kind of situation and reacts in a similar way. Each wants to change, to modify, or to convert the world represented by one state into a world represented by a different one. The initial state might be San Francisco without the Golden Gate Bridge; the final state, San Francisco with this bridge. The initial state might be the Nile without a dam; the final one, the Nile with a new dam in place. Or the initial state might be a Neanderthal contemplating the death of a loved one; the final one, the world after the construction of a sepulcher. But we can also view the maneuvering to obtain a divorce, the firing of the president of Iran, or the creation of a better bacterial host in a completely analogous way. Can we deny that the world is somehow different after the composition of a song, the painting of a picture, or the construction of a rhyme? Graphically each of these examples is represented in Figure 1, where time is given on the horizontal line or axis and some measure of change in the world is given on the vertical axis.

The engineer is to cause the transition from A to B. To identify a situation requiring an engineer, seek first a situation calling for *change*.

Yet from an engineering point of view not all changes or all final states are equally desirable. Today few would suggest replacing the Golden Gate Bridge as a means of crossing San Francisco Bay with one of the wooden covered bridges that were once commonly seen in Maine. To identify a situation calling for the engineer, second we must look for one in which not just any change, but the *best* change, is desired.

Unfortunately the engineer cannot select this best change from all conceivable transitions from the initial state to the final one. Physical, economic, political, and artistic constraints always exist. Despite its favorable corrosion properties no consideration was given to building the Golden Gate Bridge of an alloy of pure gold, for obvious reasons. In the third place, then, the engineer always seeks the best change within the available *resources*.

The final characteristic of a situation requiring the engineer is that knowledge about the system before, during, and after the transition is incomplete, inconsistent, or would require more time to accumulate than the lifetime of the problem. Again the Golden Gate Bridge will serve as a good example. Design of a bridge is a complex operation requiring knowledge of strength of materials, wind conditions, economics, politics, mathematics, fluid flow, and so forth. No engineer or team of engineers could ever hope to acquire all of the information necessary to design the perfect Golden Gate Bridge, even if such an absolute ideal were to exist. This *uncertainty*, this complexity, this doubt about the system or about the exact problem statement itself—this poorly understood situation—is the sure mark of a problem that requires an engineer to roll up his sleeves and go to work.

Now we will look in detail at the key words *change, resources, best*, and *uncertainty*, all of which have appeared in the definition of an engineering problem situation. Of these four, the reasons for including the first two are relatively easy to explain. The third is less well understood and more philosophically charged. It will occupy more of our time. Throughout the discussion of these first three key words we will sense the fourth, the lack of information or the uncertainty that always pervades an engineering problem, menacing us from the wings. After this mise-en-scène, the engineering strategy itself can make an appearance in Part II of this discussion.

Change

Engineers cause change. I know of no engineer who would disagree with that statement. Prompted by their senses, most people will agree with the engineer that change does occur and that the world is somehow different from one moment to the next. Close the book you have in your hand and reopen it. Can you deny that it was open, then closed, and now open once again? The engineer, if he thinks about it at all, finds it hard to believe that anyone could seriously question the existence of change, especially the change he causes, and dismisses anyone who professes to do so as a charlatan. An engineering colleague once told one of my classes, "I consider anyone who is not actively engaged in causing engineering change a leech on society." Fortunately for him, most Western philosophers would at least agree that change can occur, but there have been some notable exceptions. In ancient times Parmenides taught that change was impossible and felt that the changes we seem to perceive are but illusions of the senses. Surely it must seem to us in the modern age that Parmenides was delusional, and his claim unworthy of being used as an example. Is there any chance that the delusion is instead ours? More recently Santayana has held that doubt is always possible about the existence of actual change, and McTaggart has explicitly denied that change exists. The engineer may be startled to learn that Leibnitz, who in parallel with Newton invented the branch of mathematics most concerned with change—the calculus—had serious reservations about the matter. In addition, consideration of the work of Bradley and expansion of our discussion to include the view of Eastern philosophy, where denial of change is even more prevalent, show that, contrary to the belief of my engineering colleague, there are serious philosophical problems in determining if change exists and what it is. By insisting that an engineer is involved in changing the world, we have taken a philosophically risky first step. Given the equally ambiguous notion of causality in modern philosophy, by professing that engineers can cause this change, we have taken an even more precarious second one. Just for the moment, note well, we will assume that change occurs and will consider that change for which the engineer claims responsibility.

We immediately run into four practical difficulties when we consider the changes for which the engineer claims credit: The engineer doesn't know where he is, where he is going, how he is going to get there, or if anyone will care when he does. Initially, the engineer is located at point A in Figure 1. No one would claim to have perfect knowledge about the world at the time represented by this point. The second problem is that the exact final state, point B, is also unknown when the engineer first goes to work. An example will make this point clear. The Aswan High Dam in Egypt has increased the salinity of the Nile by 10 percent, has led to the collapse of the sardine industry in the Delta, has caused coastal erosion, and has forced the one hundred thousand Nubians displaced by the reservoir to try to adapt to life as farmers on the newly created arable land. These liabilities have been balanced, some would say more than offset, by other assets such as the generation of enough hydroelectric power to furnish one-half of Egypt's electrical needs, etc. Our interest, however, is not to critique this spectacular engineering project or to reconcile conflicting opinions as to its net worth, but to emphasize that before construction, at state A, the engineer could not have predicted the exact change in salinity and erosion or the exact human costs to the sardine fishermen and the Nubians. The final state always has a reality that an engineer situated at the initial state cannot anticipate. Likewise, the presidential order to "put a man on the moon by the end of the decade" lacks the specificity of the ladder Neil Armstrong descended to leave his footprint on the moon. The engineer is willing to develop a transition strategy from the present to a presumably more desirable future state, but rarely is he given a specific, well-defined problem to solve. Instead he must determine for himself what the actual problem is on the basis of society's diffused desire for change. At the beginning of an engineering project the engineer rarely knows exactly where he is going.

The third difficulty is with the change itself. Figure 1 falsifies the ease in deciding which path to take from A to B by showing only one. Usually a number of alternatives exist, each limited by different constraints. The engineer is not responsible for implementing a single given change, but for choosing the most appropriate one. In other words, at state A he doesn't know how he is going to get to state B even if he were to have complete knowledge of the problem he is to solve.

The final difficulty in causing change is that an engineering goal has a way of changing throughout a design. From the start of a project to completion is often a long time. At present, for example, it takes twelve years to construct a nuclear reactor in America. During the completion of an engineering project, changes in the final goal often occur requiring a reorientation of the project in midstream. In the automobile industry the public's demand has flitted from the desire for a powerful automobile, to a safe automobile, to a small, fuel-efficient one, and back to a larger one at a rate that has often left a new automobile design obsolete before it leaves the drawing board. With the lack of information about points A and B and the desired transition path between them, combined with changes in point B throughout the project, how can the engineer ever hope to cause the change he desires? Change is recognized as a characteristic of an engineering problem, but with all of the attendant uncertainty, what *strategy* does the engineer use to achieve it?

Resources

The second characteristic of a situation that requires the services of an engineer is that the desired solution must be consistent with the available resources. These resources are an integral part of the problem statement and both define and constrain its solution. Different resources imply different problems, and different problems require different solution techniques. To make this point, one of my former engineering professors* would begin each class period with a simple problem to be answered in fifty seconds by what an engineer would call a *back of the envelope* calculation. Once, for example, we were asked to estimate the number of Ping-Pong balls that would fit into the classroom. In addition to developing the ability to manipulate large numbers in our heads, these problems taught the importance of resources in the definition of a project. Based on fifty seconds we were to provide an answer to the problem—a correct engineering answer. Had we been given two days for our response, we would have been expected to measure the room and calculate the number—again, an entirely correct engineering answer given the additional resources. I suppose if we were given even more time and needed the best possible answer, we should have filled the room with Ping-Pong balls and counted them. Although obviously similar, each of these problems was fundamentally different as evidenced by their need for different methods of solution. The answer to each was absolutely correct from an engineering point of view when both the problem and time constraint were considered together.

Contrast an engineering problem to a scientific one with respect to their dependency on resources. Although Newton was limited in the amount of time he had to develop his theory of gravitation and a modern cancer researcher is constrained by the funds at her disposal, we usually think of each as trying to read the already written book of Nature instead of creating a new best-seller based upon the available resources. We quibble, by extending the analogy beyond its bounds, if we try to argue that Nature, and by implication science, has a correct answer to the Ping-Pong ball problem and that the engineer is only limited by his resources in his approximation to this number. A similar sense of convergence to truth does not usually exist in an actual engineering problem. For example, if we try to argue that Nature has an absolutely correct answer as to whether the Aswan High Dam should have been built and that the engineer will converge to it with additional resources, we quickly become inundated in profound philosophical water. Instead of looking for the answer to a problem, as does the scientist, the engineer seeks an answer to a problem consistent with the resources available to him. This distinction will become clearer when we later consider the engineer's notion of best. At least for now it seems reasonable to agree with the famous engineer, Theodore Von Kármán: "Scientists discover the world that exists; engineers create the world that never was."

An engineering problem does, however, require evaluation of the true resources that are available. This is difficult for three reasons. We usually think of resources as time and money because they are often explicitly stated in the problem, are usually in short supply, and are depletable. In exceptional cases other depletable resources may also be important. Since a good engineering rule of thumb is that approximately one hundred pounds of rocket are required to place

---

*Dr. J. J. McKetta, Jr.

one pound of satellite into orbit, weight is often limited in space exploration, and the engineer treats it as a precious resource to be wisely managed.

But depletable quantities such as time, money, and weight are only one part of what is meant by a resource in an engineering context. An engineering resource is anything that makes an important or significant difference to a project. To determine if something should be considered as a resource, perform this simple test: Imagine two teams of engineers identical in all respects but one. If the final product of one team is judged preferable to the product of the other, then the difference between the two teams should be taken as a resource. As an example, consider two engineering teams with the same number of members, same education, and same amounts of time and money at their disposal. One team, however, is more experienced solving problems similar to the one we pose. Past experience with similar problems usually produces a better design. *Past experience with similar problems* should, therefore, be considered as a resource, although obviously it is not a depletable one. As a second example, consider the project of writing a book. Money and time to aid in the endeavor are clearly resources, but what of *interest* and *encouragement*? I believe the author given these last two effectively has been given a resource, and both the speed of completion and the quality of the writing are enhanced. Again, the important point is that neither *interest* nor *encouragement* is depletable. By a similar analysis, *team compatibility, organizational ability,* and *enthusiasm* should also be considered as important resources. The first difficulty in determining what should count as a resource is focusing exclusively on depletable resources.

A second difficulty is confusion between the nominal resources (those explicitly named in the problem statement) and the actual resources. Perhaps a problem is nominally supposed to be finished within a certain budget or within a fixed time period, but common practice allows a 10 percent overrun in either. The design proposal that is unaware of this fact is inaccurate and, in competitive bidding, at a definite disadvantage. Large government projects are especially prone to this effect. Another example is the management of funds for research into a serious disease, such as cancer. Instead of using all of the available money on research, common practice is to use some of the money for a fundraising drive. In this way the actual resources for research may be many times greater than the original nominal ones.

The final error made in evaluating the resources available to a project is neglecting the efficiency of their use. Although the total nominal resources are fixed, allocation within this total is more flexible. Given two design teams, the one that most strategically allocates its resources or converts one kind into another at the appropriate moment has effectively increased the amount available. Money may be used to assemble a more experienced design team or to buy more equipment. Time may be allocated to search the literature for relevant information or to increase the basic research to obtain this information. A good allocation strategy effectively increases the available resources. In addition, one resource may be converted into another. For instance, hiring additional people at certain steps in a project effectively exchanges money for personnel. But if completion of this step is critical, in the sense that failure to complete it blocks further progress and forces other people and machinery to remain idle, the increased expense on personnel will result in an overall saving. The coffee break is a curious example of the exchange rate between resources. Here a small amount of time is purposely given up for an increase in enthusiasm and presumably overall productivity of the work force. When the flexibility of allocating resources and exchanging one for another is decreased, the design often suffers. In the project to put a man on the moon, the deadlines within the project were set, in large measure, by political considerations, and, hence, were relatively inflexible. As a result preliminary suborbital and orbital missions were scheduled so closely together that time did not permit the results from one experiment to be analyzed and used in the design of the following one. Instead, results from one experiment could only affect later missions in a leapfrog manner. Putting a man on the moon in a long period of time as opposed to "by the end of the decade," although unacceptable politically, would have been far less costly. The efficiency with which resources are used is as important a consideration as the total amount in defining and limiting an engineering project.

An engineering problem is defined and limited by its resources, but the true resources must be considered. Because we tend to think only in terms of depletable resources, because we confuse nominal and actual resources, and because we neglect the efficiency of allocating resources and the possibility of exchanging one kind for another, often the true resources are hard to determine. Given the uncertainty in determining what should be considered a resource, how can the engineer ever hope to provide an adequate solution to a problem? What *strategy* does he use?

Best

We have agreed to consider the four characteristics of an engineering problem captured in the keywords *change, resources, best,* and *uncertainty.* The first two of these were easy to understand. We can easily accept that an engineer wants *change* and, with some thought, almost as easily accept that the definition of an engineering problem depends on the available *resources.* The third characteristic of a problem that requires the services of an engineer, the appropriate notion of *best,* is more of a challenge to understand and must occupy more of our time.

The engineer wants the best solution to his problem, but then who wouldn't? The difficulty is that the engineer's notion of best represents a surprising departure from the standard concept as endorsed by most Western philosophers and as understood by most nonengineers. Most of us would accept Plato's notion (without recognizing to whom we owed the debt) of an ideal, perfect form of, say, beauty, justice, or whatever as an ultimate best and then consider approximations to this form as better and better as they approach this ideal. In fact, it is even difficult for the philosopher to conceive of an alternative to Plato's linear progression toward the ideal form if she has not studied engineering. Since our ultimate goal is to discover a universal method by generalizing the method used by the engineer, we must have a very clear understanding of what the engineer means by his terms before we proceed. This is especially critical when the engineer's definitions are at odds with both common and accepted philosophical practice. What we now seek, therefore, is the engineer's notion of a best solution or what is technically called the *optimum* solution. What we will find is a new, radical concept of *best* little used in Western, Greek-based philosophy. This adventure will carry us into the heart of an important area of engineering called *optimization theory,* which is familiar, at least in outline, to all engineers but relatively unheard of and more challenging to the philosopher and nonengineer.

This investigation will be worthwhile because, in addition to defining the engineer's notion of best, it will indicate, once again, the crucial lack of information that is always present when an engineer solves a problem and with which his method must cope. After some preliminaries, the discussion will be divided into three parts: a simple example to explain the most important technical terms used in optimization theory, a short theoretical analysis, and finally some practical considerations. Mentioning optimization theory is not meant to imply that all engineers use it formally in all projects. Instead it is included at this point to capture the essence of the engineer's notion of best.

Best is an adjective applied redundantly to describe an existing engineering design. That a specific automobile exists proves that it is some engineer's subjective notion of the best solution to the problem he was given to solve. Saying that a Mercedes is a better automobile than a Mustang is nonsensical if *better* is being used in an engineering sense. They are both optimum solutions to different specific design projects. Likewise, the complaint that "American engineers cannot build an automobile that will last for fifty years" can only be voiced by a person with little understanding of engineering. To construct such an automobile is well within the ability of modern automotive engineers, but to do so is a different design problem from the one currently given to the American engineer. It does make sense to prefer one design project over the other. An engineer could conceivably argue that designing an automobile similar to the Mercedes is a better goal than designing one similar to the Mustang because it would last longer, conserve natural resources, promote national pride, or whatever. And, of course, a second engineer may feel that he could have produced a better final product than the first engineer given the same problem statement. But for the engineer who designed the Mustang, the automobile you see before you is his best solution to the problem he was given to solve. *To exist is to be some engineer's notion of best.*

Defining the technical terms used in optimization theory will help to make this point, although the discussion that follows may be somewhat tedious for the nonengineer. Still, it should be carefully studied for it has profound philosophical consequences for the definition of universal method.

Consider a television set with one control. We will assume that turning this knob to a higher number will produce a better picture but at the same time worsen the sound; turning the control to a lower number, on the other hand, will worsen the picture but improve the sound. Confronted with such a device, it would be relatively simple for you to adjust it for your personal preference as you balance the relative importance of picture and sound to you. The engineer's job, however, is not only to please you, but also to find the best permanent adjustment of this knob to please society. Let us see how this is done theoretically. The setting of the knob is called a *manipulated variable*. One can set it to a lower or higher number as desired. The quality of the picture (say, sharpness) and the quality of the sound (say, fidelity) are the *criteria* in the problem. It is against these two characteristics that a judgment is made as to whether or not the control setting is best. In general, the criteria are conflicting in that an improvement in one implies a worsening of the other. The criteria taken together make up the *optimization space* or *axis system* of the problem. Figure 2 is a graph that shows how the sharpness of the picture might change with the position of the knob. As the number on the horizontal axis increases, the sharpness, shown on the vertical axis, is improved. The dotted line indicates that a setting of 4 corre-

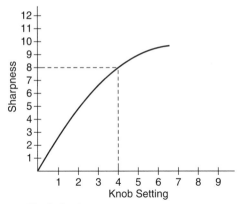

Figure 2 Sharpness vs. Knob Setting

sponds to a sharpness of 8. The sharpness of the picture is higher for a setting of 6 than for one of 2. For no compelling reason this curve is often called a *return function*.

A different return function, Figure 3, exists for the sound. Turning the knob to a higher setting decreases the fidelity of the sound. When the control knob is set to the value 4, the value of the fidelity is also 4, as can be seen on the graph.

To represent pictorially what a person does instinctively when he selects his preferred setting, the two return functions must be combined or superimposed on the same graph; they must be put on the same basis; or, technically, they must be made commensurate. This requires that a *common measure of goodness* be found for picture and sound. What the engineer seeks is the relative importance of sound and picture to the owner of the television set. In other words, a 10 percent increase in the sharpness of the picture is worth what percentage decrease in the fidelity of the sound? This relative importance is expressed by the *weighting coefficients* of the two conflicting criteria. Another term used for the set of weighting coefficients is the *value system* of the problem.

Let us assume for the moment that sharpness and fidelity are equally desirable. That is, the relative *weights* in the two cases are equal. The resulting com-

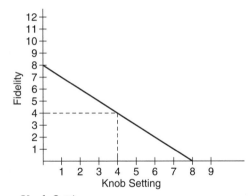

Figure 3 Fidelity vs. Knob Setting

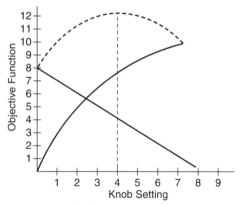

Figure 4 Objective Function vs. Setting

bined graph is given in Figure 4. The bottom two curves are the return functions for the conflicting criteria that were shown in Figure 2 and Figure 3. The upper dotted curve is the sum of these two under the assumption that improvements in both picture and sound are equally desirable. Verify that this combined dashed curve with a value of 12 at setting 4 is equal to our previous reading of 8 and 4 at the same point. This dashed curve is the common measure of goodness for the problem. It is sometimes called the *objective function* or *measure of system effectiveness*. We want the largest or maximum value of the objective function. This number is called the *optimum* or *best value*. For our television set the optimum setting corresponds to a control setting of the knob or manipulated variable of 4.

Optimization is a compromising process. If the setting is below 4, the sound is better and the picture worse, but in such a way that our net satisfaction with the combination is lower. Likewise, a setting above 4 produces a better picture but this is more than compensated for by a reduction in the fidelity of the sound, and again our satisfaction is less. From an engineering point of view point 4 is the *best*, or optimal, setting. The engineer calls the process of balancing an improvement in one criterion against a worsening of another a *trade-off*. One cannot have the best of all possible worlds—in this case the world of sight and the world of hearing. The most one can hope for is the best in the combined, real world.

Although the preceding example is simple, it contains the important theoretical aspects of optimization theory. If the engineer can develop a mathematical model for his problem including the appropriate optimization space, return functions, and commensuration, the optimum is determined. This does not mean that it is easy to find, but only that in principle it can be found. The design of an automobile will have an optimization space with a much larger number of conflicting criteria such as fuel economy, power, safety, comfort, and style, and it will have more complicated return functions. As a result, a large computer and sophisticated numerical strategies would be needed to obtain the optimum if we wanted to calculate it analytically. But the important point is that given a valid model, evaluation of *best* is a well-defined engineering operation.

Unlike science, engineering does not seek to model an assumed, external, immutable reality, but society's perception of reality including its myths and prejudices. If a nation feels that a funeral pyre should be aligned in a north-south direction to aid the dead's journey into heaven, the model to be optimized will incorporate this consideration as a design criterion irrespective of the truth of the claim. Likewise, the engineering model is not based on an eternal or absolute value system, but on the one thought to represent a specific society. In a society of cannibals, the engineer will try to design the most efficient kettle. As a result, the optimum obtained from this model does not pretend to be the absolute best in the sense of Plato, but only the best relative to the society to which it applies. Contrast this with a scientific model. Speaking of Einstein's theory as the best available analysis of time and space implies that it comes closest to describing reality than any alternative proposal. It is considered better than the formulation of Newton because it explains more accurately or more simply our observations of an assumed, external, immutable Nature—a "Nature behind the veil." To emphasize the difference in the engineer's and the scientist's notions of best, consider the odd situation that would arise if time and space turned out to be human-made myths. I grant that for most of us this is an unbelievable situation, but some philosophers believe just that. For example, in the book *The Tao of Physics* by Fritjof Capra, Swami Vivekanada is quoted as saying, "Time, space, and causation are the glass through which the Absolute is seen. . . . In the Absolute there is neither time, space or causation." With no external time and space to refer to, the scientist's *best*, as an approximation to these concepts, becomes vacuous. On the other hand, the engineer's *best* is still a valid model of society's (now) mythical concept of time and space. *Best* for the scientist implies congruence with an assumed external Nature; *best* for the engineer implies congruence with a specific view of Nature.

The appropriate view of Nature for optimization by the engineer is not just an objective, faithful model of society's view, but includes criteria known only to him. One important consideration in lowering the cost of an automobile, for instance, is its *ease of manufacture*. If standard parts can be used or if the automobile can be constructed on an assembly line instead of by hand, the cost of each unit goes down. *Ease of manufacture* is a criterion seldom considered by the public when it decides which automobile to buy but one that is essential to an accurate model for determining best automobile design. To quote a prominent engineer Karl Willenbrock from an editorial in *Science* magazine, "Their [engineers'] designs must satisfy scientific as well as nonscientific criteria such as manufacturability, maintainability, risk-minimization, and cost-effectiveness." Because of these additional variables, the appropriate optimization model is not just a surrogate for society but includes subjective considerations of the engineer who makes the design.

In general, the optimum is different when an optimization space with a reduced number of criteria is used. The best automobile from the points of view of the public and the engineer will therefore differ. The person who criticizes the engineer for not providing an automobile to last fifty years makes the error of not using a complete axis system. She is almost certainly not considering the ease with which such an automobile could be manufactured, and so forth. Her view of the true optimum in this reduced axis system is as deficient as the knowledge of a three-dimensional world to the two-dimensional inhabitants confined to live in a two-dimensional world in Abbot's classic novel *Flatland*. The design of a long-life automobile is possible and would be as exciting a challenge to the engineer as the present line of products. But the demand for one in the United

States is so low, and the cost of producing it so high, that the cost per vehicle would be prohibitive. By the way, such a car, the Rolls-Royce, is actually on today's market, but if all automobiles in America cost as much as a Rolls-Royce, few Americans would need a two-car garage.

As a second example of a deficient system of axes being used by some members of the public, consider the complaint communication engineers occasionally hear, "This holiday season all of the telephone lines were busy and I couldn't get through. You would think the people at the telephone company could anticipate the rush." Again we have two different axis systems being used—one by the layperson and a different one by the communication engineer. The engineer could easily design a telephone system for the busiest period of the year, but the extra equipment that would be needed would remain idle the rest of the year and would have to be stored, insured, and maintained. The engineer uses an axis along which the cost of the extra, seldom-used equipment is being traded off against the loss of service. I agree that the public has a right (I would say obligation) to help select the problems for solution, the major criteria to be included in the design, and the relative importance of these criteria. The engineer, however, always uses additional criteria unknown to the public. Since in general the optimum will shift depending on the number of variables used, it is naïve for the layperson to criticize the engineer's final design without justifying why the reduced set was more appropriate.

Theoretically, then, *best* for an engineer is the result of manipulating a model of society's perceived reality including additional subjective considerations known only to the engineer constructing the model. In essence, the engineer creates what he thinks an informed society should want based on his knowledge of what an uninformed society thinks it wants.

Although the theoretical procedure described here is useful for indicating what an engineer means by *best*, it runs into serious practical difficulties. Accurate knowledge of the optimization space appropriate for society is often unavailable, and, even if it is available, it does not necessarily reflect the desire of any one individual. Society is an abstraction for a group of individuals each of whom has her own personal definition of best. This abstraction is necessary because a specific television set is to be produced, which implies one optimization space, one set of return functions, and one set of weighting coefficients. If we were to ask different people to adjust the television set that was used in the earlier example each would choose a different setting of the knob. A person who was hard of hearing or interested in high fidelity would place more emphasis on sound than picture, and her optimum setting would shift.

Figure 5 is drawn with the quality of the picture half as important as that of the sound. For convenience the return function for the sharpness of the picture has been redrawn reduced by one-half so that the combined curve can be calculated with ease. Verify that the value of the objective function at a setting of 4 is now equal to 8, or one-half of the old value for sharpness (4) plus the old value for fidelity (also 4). As expected the optimum has now shifted to a lower setting—one that favors sound. Each person will choose a different value depending upon the relative importance of sound and picture to her. In this example it would be easy to determine the best setting for each individual experimentally by simply asking her to set the dial as she desires. But in actual practice the engineer must determine what best means for each individual before the product is even built. Therefore, he must guess what criteria each individual considers important and the relative importance attached to each before con-

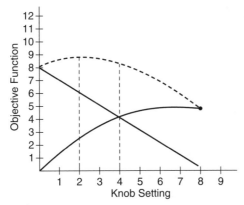

Figure 5 Revised Weighting

struction is begun. From a practical point of view neither can be determined perfectly. Assume, however, that somehow we know these quantities. To use our theoretical optimization procedure, the separate optimization spaces for all of the individuals that make up society must now be combined to produce a representative average. No unique, completely defensible average exists. It will depend on whether a numerical, political, or economic scheme is used to weight the individual optimization spaces to produce it. In practice the engineer must determine this average. Assume, however, that this has also been done. At last we are able to evaluate the theoretical best for society, but unfortunately we now have no assurance that this average is acceptable to any one member of the group. If one-half of the public wanted a vehicle with four wheels and the other half preferred one with two, I doubt the engineer should design a three-wheeled car. Or as one of my former chemical engineering professors* put it memorably, although a bit earthily, "Always remember that the engineer designs for the individual, not the average. The average American would have one tit and one ball."

Although the theoretical procedure for determining the best design is well defined, the practical procedure is not. In some completed engineering projects we have experimental evidence that the axis system ultimately chosen as representative of society was deficient. The San Francisco Embarcadero, originally begun in 1959, has become a classic example of the practical problem of trying to evaluate society's optimum. It was designed as the best way to move traffic about the city. Money was appropriated and construction was begun. But the Embarcadero is now known as the "freeway that goes nowhere."

The reason is shown in Figure 6, where we see the end of an uncompleted ramp. It was abandoned in mid-construction because the design failed to include considerations that ultimately proved important. Criteria such as "Don't increase the noise level or density of people in my neighborhood" and "Don't decrease the overall quality of life" were important to the citizens of San Francisco. Too expensive to tear down, the Embarcadero stood for many years as a monument to the difference between engineering theory and the real world. To quote an article in the *Los Angeles Times* (November 5, 1985):

---

*Prof. Van Winkle.

Figure 6 San Francisco Embarcadero

> Politicians here have for 20 years won hearts—and votes—by pledging to de-
> molish an ugly, unfinished stump of the elevated, double-deck Embarcadero
> Freeway along the city's waterfront. On Monday, the Board of Supervisors
> voted 8 to 2 to proceed with preliminary engineering on a project that would
> exchange the freeway for a tree-lined boulevard and better mass transit. . . .
> For years, the freeways—especially the Embarcadero—were seen as useless
> eyesores that went nowhere and blocked views.

Evidently, the $10 million axis system left out a criterion that was important to
the voters.

As an additional example, remember the urban renewal projects in New York
City that were designed to be optimal with respect to money, space utilization,
and materials. They proved to be suboptimal when the tenants broke windows,
dumped garbage in the halls, and moved out.

Figure 7 serves to introduce a third and final example of an engineering project that turned out to be suboptimal. It depicts the schematic of the floor plan
used in two skyscrapers built on the campus of a well-known (but charitably
unnamed) university in the North. Beginning at the center of the figure, we note
the elevator and utility areas surrounded by six living areas. Each living area
consists of a bathroom and living room for four separate suites. Each suite has
a study area with four desks for the four inhabitants who sleep in bunk beds in
the outer area. In all, sixteen people are housed in each of the six living areas.
It is also worth noting that none of the walls meet in right angles. Psychologists
tell us that Westerners become disoriented when they are in rooms with odd

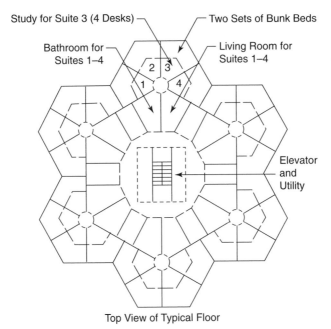

Study for Suite 3 (4 Desks)

Two Sets of Bunk Beds

Bathroom for
Suites 1–4

Living Room for
Suites 1–4

Elevator
and
Utility

Top View of Typical Floor

Figure 7 Mystery Floor Plan

corners. In addition, the windows on the extreme outside were reduced to tiny slits similar to those used in castles to protect archers defending the fortress. This innovation was evidently made to minimize the air conditioning cost by reducing the window area. The overall floor plan looks strikingly like a beehive, and well it might. The beehive has evolved over centuries to have maximum capacity (for honey) for a minimum amount of construction materials (since bees have to transport the construction materials in many trips.)

The history of these buildings is revealing and serves as experimental evidence that engineers do not always get the optimization space right. Originally these skyscrapers were to serve as modern, desirable, state-of-the-art dormitories, but when students refused to live there by choice, they had to be made into required freshmen dorms. I have been told that when even this did not produce a satisfactory occupancy rate, with poetic justice, at least one of them was made into an administrative building. These buildings were simply not optimal as demonstrated by the exodus of students. Although theoretically the best design is determined once the optimization space is known, practically it is hard to be sure that we have not neglected an important axis in constructing this space. In all of these three examples, the optimization space of the engineer proved in practice to be a poor representation of the one preferred by society.

Please remember the engineer's notion of best or optimum well. By the time we have completed Part IV it should become obvious that it is preferable to the classical Platonic one in defining the universal method. *The best we can do is not the best; the best we can do is the best we can do.*

A fundamental characteristic of an engineering solution is that it is the best available from the point of view of a specific engineer. If this engineer knew the absolute good, he would do that good. Failing that, he calculates his best based on his subjective estimate of an informed society's perception of the good. With doubt about the criteria that are important to society, with doubt about the relative importance of these criteria, and with doubt as to whether society's best reflects the individual's best, how can the engineer design the optimum product? What *strategy* does he use?

Uncertainty

Not much needs to be said about the last characteristic of an engineering problem, *uncertainty*, because we have seen its specter pervade each of the previous three characteristics. Indeed this uncertainty, this doubt, this struggle to solve a problem in the absence of complete information is one of the identifying signatures of an engineering problem. Until someone can remove this uncertainty, I, along with my engineering colleagues, must continue our struggle to cause the best change in an uncertain situation within the available resources. What *strategy* should we use?

EXAMPLE ENGINEERING PROBLEMS

Best, change, uncertainty, and *resources*—although we do not as yet know what the engineer's strategy for causing change is, it should not be too difficult to recognize a situation calling for its use. But unfortunately it is. When President Reagan promoted a new generation of space weapons to create a defensive umbrella for America and then called on the *"scientific* community" to give us a way of developing it, he confused science and engineering. Relatively speaking, little new science is involved. Newton's law of gravitation, the equations of motion, and the theory of energy emission by lasers or particle beams are all reasonably well understood by the scientist. If such a device is to be developed, the president would have been better advised to call on the *"engineering* community." Journalists share this same confusion about what constitutes a scientific problem and what constitutes an engineering one. When reporters seeking information about Reagan's project went to "scientific experts" to evaluate the "feasibility of this space-age missile defense system," they went to the wrong place. Its feasibility is certainly more in doubt because materials able to survive the tensile stresses, radiation damage, and alien temperature environment or because a device able to maintain the severe pointing accuracy needed cannot be found than it is because something that violates the known laws of Nature will be found. If feasibility is the question, journalists should contact the local dean of a college of engineering, not their resident scientist.

Since confusion evidently exists in the mind of the nonengineer as to what constitutes an engineering problem, let us consider several additional examples with the defining characteristics of one in mind.

The statement of an engineering problem might well be:

> I believe that this nation should commit itself to achieving the goal, before this decade is out, of landing a man on the moon and returning him safely to the earth. No single space project in this period will be more impressive to mankind, or more important for the long-range exploration of space; and none so difficult or expensive to accomplish. . . . [The cost would be] $531 million in 1962 and an estimated $7–9 billion over the next five years.

President Kennedy's exhilarating problem statement before Congress was the gun that sounded the start of one of America's most spectacular engineering races. Make no mistake: It was primarily a race between engineers, not scientists.

Similarly, when Congress passed and President Johnson signed the Appalachian Regional Development Act, an engineering problem was implied:

> This bill authorized a total of $840 million in federal grants over the six-year period ending June 30, 1971, to pay up to 70 percent of the cost of building 2,350 miles of development highways and 1,000 miles of local access roads in the region.

Reagan, Kennedy, Johnson—in these three examples a president asked for the best change from an initial state A to a different, more desirable state B for a complex, poorly understood system within the resources at his disposal. For each, he needed the services of people who understood the engineering method.

Engineering change is not limited to the creation of physical devices such as defense systems, spaceships, or highways. Consider the following statement:

> We . . . have already declared war on poverty in all its forms. . . . We have $8,700,000 going into vocational training of 3,000 persons for 50 occupations. We have $4 million in loans and grants for housing. We have surplus food distribution in all the eastern Kentucky counties. We have 360 schools, 10,000 pupils, participating in the school lunch program. Under our community work and training program, we have 48 programs going, 1,400 men in action.

Once again, we sense the concepts *best, change, uncertainty,* and *resources* all used as an engineer would use them; once again, President Johnson should have called on the engineering community for help.

Other political, economic, and psychological examples that require the talents of the engineer are easy to find. Some are almost trivial, others are more complex—but all have the characteristics of an engineering problem. Perhaps a politician wants to be reelected or to win support in Congress for the construction of a dam in her home district; perhaps an economist would like to increase the gross national product or find a way to reduce the national debt; perhaps a psychologist would like to stop a child from biting her nails or condition a race to create a utopian state using behavioral engineering. The changes implied by these examples are usually not associated with the engineer, but careful study of the characteristics they share with the obvious engineering projects of designing a nylon plant, constructing a bridge across the Mississippi River, and building an electrical power station for New York City shows a definite pattern. For each, the engineering method is needed.

If you, as with all humans since the birth of humankind, desire change; if the system you want to change is complex and poorly understood; if the change you will accept must be the best available as you balance often conflicting criteria; and if it is constrained by limited resources, then you are in the presence of an engineering problem. If you cause this change using the strategy described next, then you are an engineer.

The Principal Rule
of the Engineering Method

I was, during the spring of 1965, in graduate school in the Northeast, and as the weather was pleasant, not cold as it sometimes is in Boston, the window of the spacious room I shared with another engineering student was open to the famous New England sunshine. We had some time for leisure, and since my roommate* was one of the highest ranked chess players in the United States, he proposed to teach me the game. I should make it clear at the outset that although I knew enough not to confuse knights with bishops and by sustained effort could move each piece properly, I really have neither the temperament nor the aptitude to play chess. I was, as my roommate constantly reminded me, a complete patzer when it came to the game.

Although chess is a complicated game, in theory, perfect play is possible. In principle, what an engineer would call a complete game tree can be constructed for the game of chess. By exhaustive enumeration of all the legal first moves by white, followed by all of the legal responses by black to each of white's moves, then white's new responses to each of black's moves, and so forth, every possible game can be configured as a tree. Next a well-defined operation called *min-maxing* the game tree from the bottom up would allow the perfect game to be identified. This procedure is well understood and guaranteed to be successful. Even the smallest computer can be programmed to play perfectly the much simpler game of tick-tac-toe using this strategy. In practice min-maxing the game tree is impractical for chess, however, because of the enormous number of different positions[†] and the limited resources of even the largest computer. Chess, therefore, defies analytical analysis. Yet people do play chess, and as I wanted to be counted among them, something had to be tried.

My roommate used a different strategy to cause *desirable change* in my *poor understanding* of chess consistent with the *available resources*. This strategy consisted of giving me suggestions, hints, and rules of thumb for sound play. He began by telling me:

---

*Dr. Carl Wagner.

[†]This number is estimated as well over 10^{120}, which is more than the number of grains of sand on all of the world's beaches.

1. Open with a center pawn,
2. Move a piece only once in the opening,
3. Develop the pieces quickly,
4. Castle on the king's side as soon as possible, and
5. Develop the queen late.

As I got better, I began to hear:

6. Control the center,
7. Establish outposts for the knights,
8. Keep bishops on open diagonals, and
9. Increase your mobility.

Throughout my lessons I had a curious feeling. Although my roommate's hints did not guarantee that I would win, although they often offered conflicting advice, although they depended on context and changed in time, I did learn to play better chess much more quickly than I would have learned it had I tried to create the game tree. What was the word for this strange *strategy* to cause a change in my ability to play chess?

At about this time,* I was taking a course in artificial intelligence for my minor, in which we studied an unusual way of programming a computer to solve problems. Instead of giving the computer a program with a fixed sequence of deterministic steps to follow, an algorithm as it is called, it was given a list of random suggestions, hints, or rules of thumb to use in seeking the solution to a problem. These hints were called *heuristics;* the use of these heuristics, *heuristic programming.* Surprisingly this vague, nonanalytic technique works. It has since been used in computer codes that play championship checkers, identify hurricane cloud formations, and control nuclear reactors. By a chance coincidence of this course in artificial intelligence, of the desire to learn chess, and of a personal crisis (to be described later) in completing a mathematics assignment, I found the name of the *strategy* we were constantly seeking in Part I. Both the method for solving my problem (learning to play chess) and that of the engineer in solving his problems (building bridges, dams, and so forth) depend on the same strategy for causing change. This common strategy is the *use of heuristics.* In the case of the engineer, it is given the name *engineering design.*

To analyze the important relationship between engineering design and the heuristic, five major objectives are set for this part of our discussion. They are to give a preliminary definition of engineering method; to understand the technical term *heuristic;* to develop the engineer's strategy for change; to define a second technical term, *state of the art;* and to state the principal rule for implementing the engineering method.

One caution must be observed before we begin. The material presented here is absolutely essential for achieving the objective of this book, but the casual reader may find that it is overly tedious and too specific to the engineer. I challenge the attentive philosopher to anticipate the generalization of the engineering method to the universal method that is to come.

---

*Spring 1965.

DEFINITION OF ENGINEERING DESIGN

Engineering design is the essence of engineering. I do not wish to detract from the considerable contributions of the production and sales engineers, but design is the unique, essential core of the human activity called engineering. But for it, the engineer would not exist. We have now found what we have long sought—a preliminary definition of engineering method. Engineering design, or

> the engineering method, is the use of *heuristics* to cause the best change in a poorly understood situation within the available resources.

We have already carefully considered all of the important concepts this definition contains with the exception of the word *heuristic*. Our attention must now turn to it.

THE HEURISTIC

The heuristic will be considered by defining it, by examining its characteristics or signatures, by considering its synonyms, and by looking at several concrete examples. A more extensive list of engineering heuristics will be deferred until Part III.

Definition

An excellent way to begin would be to simply define a heuristic. From an engineering point of view this approach would be straightforward and uncontroversial. But just as we have seen earlier with the important words *classification* and *change*, from a philosophical point of view we are engaged in risky business when we try to define anything. Philosophers do not know what sort of thing a definition is, what standards it should satisfy, or what kind of knowledge it conveys. Does a definition simply say what a word means, is it an act of the mind that formulates the essence of a thing, or is it just a statement that contains the genus and differentia of a thing? Maybe a definition gives the cause of a thing or simply gives the purpose or interests we have in mind when we classify things to suit ourselves. Are we to listen to Hobbes, Aquinas, Aristotle, Spinoza, or James? But I wander into the realm of the philosopher. Until philosophers decide among themselves how to define *definition*, I intend to use mine only to guide, to discover, and to reveal—dare I say it, only as a heuristic. For that is what a heuristic is.

> A *heuristic* is anything that provides a plausible aid or direction in the solution of a problem but is in the final analysis unjustified, incapable of justification, and potentially fallible.

The exact words in this definition are unimportant, and various definitions of the heuristic exist in the literature. I have chosen one. We do need to constantly keep in mind its two characteristics: It is plausible, but it is fallible. Using this statement as a first approximation to a definition of the heuristic, or as the engineer would say, as a *first cut* at the problem, this entire book should be viewed as an aid to defining the heuristic for yourself by successive approximations.

Signatures of the Heuristic

Although difficult to define, a heuristic has four definite signatures that make it easy to recognize:

1. A heuristic does not guarantee a solution,
2. It may contradict other heuristics,
3. It reduces the search time for solving a problem, and
4. Its acceptance depends on the immediate context instead of on an absolute standard.

Since a good rule of thumb is to move in small steps from the known to the unknown, let us compare the presumably well-known concept of a scientific law with the less well-known concept of the heuristic with respect to these four signatures. Incidentally we will come to appreciate the rationality of using irrational methods to solve problems.

This comparison will be easier if we use the mathematical concept of a *set*. A *set*, sometimes called a collection or class, is an aggregate of elements considered as a whole. The set of pages in this book is a single object that has a number of members or elements equal to the number of actual pages in this book. A set may be partitioned or divided into subsets. One subset of the set of pages of this book is the set of pages that have the word *heuristic* written on them. We will only use the concept of a set in an informal way to help understand the heuristic. This is fortunate, because, although set theory was once thought to provide the definitive, indisputable, absolutely true basis of mathematics and logic, later studies* have cast doubt on this claim. The following model is admittedly simplistic and omits considerations that will become important later. As a result, it should only be used as a temporary tactic to illustrate the difference between a scientific theory and a heuristic.

The interior of the dotted rectangle in Figure 8 will be taken to represent the set of all problems that could, in principle, ultimately be solved by humankind. It will be given the name U. U is not limited to those problems that are solvable on the basis of present knowledge, but includes all problems that are theoretically solvable given perfect knowledge and an infinite amount of time. The points labeled *a*, *b*, *c*, *d*, *g*, and *h* are elements of U and represent some of these solvable problems. If you prefer, it is sufficient for our present purposes to think of U as a simple list of questions about Nature that, in theory, humankind could some day be able to answer. On this list most people born into the Western tradition would include: Will the sun rise tomorrow? Does bread nourish? If I release this ball, will it fall? Many engineers would also include the question, Should the Aswan High Dam have been built? within this area. Outside of this

---

*For instance, the peculiar properties of infinite sets, the axiom of choice, noncantorial set theory, and Gödel's Proof have led to unsolvable paradoxes in the definition of a set.

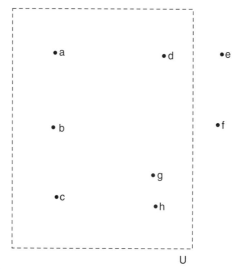

Figure 8 Set of All Solvable Problems

area is everything else: questions that cannot be answered by humans, questions that cannot even be asked by humans, and pseudo questions. The most radical scientists believe that no points such as *e* and *f* exist in this outside region. This picture admittedly leaves unidentified and certainly unresolved many important issues. Inside or outside of the dotted line, where do we put the following questions: Does God exist? Does a golden mountain exist? What is *classification*? What is *change*? What is *definition*? What is a *set*? Philosophers have fought over these concerns for centuries, and the battles have been bloody. Rather than detour now to gawk at the carnage, I propose to treat the set of all answerable questions as a faltering step down the road to our own philosophical struggle.

Figure 9 is identical to the previous one except that the sample problems are encircled by closed curves labeled A–I. Similar to the dotted rectangle, the area inside each curve represents a set. Those that have been crosshatched, A, B, B′, C, D, and I, are sets of problems that may be solved using a specific scientific or mathematical theory, principle, or law. Set A, with problem *a* as a representative element, might be the collection of all problems solvable using the law of conservation of mass-energy, and set B, those requiring the associative law of mathematics. If the area inside of a curve is not crosshatched, such as E, F, G, and H, it represents the set of all problems that may be attacked, but not necessarily solved, using a specific heuristic. This figure will help illustrate the difference between a scientific law and a heuristic based on the four signatures given earlier.

First, heuristics do not guarantee a solution. The hints available to me when I was learning to play chess certainly did not ensure that I would be able to beat my roommate. Indeed, I was able to win only one time and then only because he gave me the advantage of a rook. To symbolize this characteristic, the sets referring to scientific laws rest completely within the large set U while those referring to heuristics include area both inside and outside of U. When heuristic E is applied to problem *d*, a satisfactory solution results. This is not the case

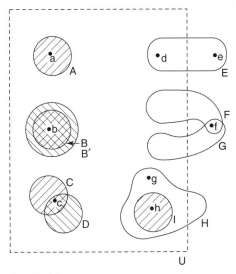

Figure 9 Representative Problems

when the same heuristic is applied to problem *e*. In this case, we think we have found a good solution to problem *e*, but this turns out not to be the case. A scientist considers ambiguity in knowing if an answer to a question has been found a fatal weakness. She seeks procedures, strategies, and algorithms that give predictable results known to be true. Uncertainty about a solution's validity is a sure mark of the use of a heuristic.

Unlike scientific theories, two heuristics may contradict or give different answers to the same question and still be useful. This blatant disregard for the classical law of contradiction is the second sure signature of the heuristic. In Figure 9 the interpretations of the overlap of the two sets that represent the scientific approach to solving problems and the overlap that represents the heuristic approach to solving problems have different meanings. In the figure, the overlap of the two scientific sets, C and D, indicate that a problem in the common area such as *c* would require two theories for its solution. The need for both the law of gravitation and the law of light propagation to predict an eclipse is a good illustration. Since combinations of two, three, and even more scientific and mathematical theories must work together to solve most problems, large areas in U are tiled with complex arrays of overlaid scientific sets. This is not true in the case of the heuristic. Here the overlap of F and G represents the conflicting answers given to problem *f* found outside of U. Although at times two heuristics might be needed to arrive at an answer and, hence, overlap within U, the most significant characteristic of a heuristic is its rugged individualism and tendency to clash with its colleagues. To answer the question, "What should I do with my bishop?" in a game of chess, one heuristic might suggest, "Move your bishop forward to an open diagonal" and another, "Retreat your bishop to protect the king." Or to consider an example closer to an engineering context, we have already seen that at least three different heuristic strategies are available to arrive at the number of Ping-Pong balls in a room and that each leads to a different, but completely acceptable, engineering answer. For a mathematician, contradiction is worse than ambiguity. She might tolerate a heuristic strategy that indicates the direction to a solution if independent confirmation exists that the solution, once found, represents the truth. A contradiction, however, is always unacceptable, for it implies a complete breakdown in her system. Logically from any two propositions that contradict, any proposition at all may be proven to be true—certainly a bothersome situation in science and mathematics. Unlike scientific laws, heuristics have never taken kindly to the harness of conventional logic systems and may be recognized when they bridle.

Some problems are so serious and the appropriate scientific techniques to solve them either nonexistent or so time-consuming that a heuristic solution is preferable to none at all. Problem g in Figure 9 is not a member of any cross-hatched set, but is a member of the heuristic set H. If g is lethal to the human species on a time scale shorter than scientific theory can be developed to solve it, the only rational course is to use the irrational heuristic method. Problem h represents a variant of this situation. It is a member of both H and I, but now let us assume that the time needed to implement the known, rigorous solution is longer than the lifetime of the problem. Again, better first aid in the field than a patient dead on arrival at the hospital.

Unfortunately, most serious problems facing mankind are similar to g and h. Sufficient scientific theory or enough time to implement known theory does not exist to solve the problems of war, energy, hunger, and pollution. But in each case first aid in the form of heuristics is surely available if only we knew how to use it. A paraphrase of Descartes gives us philosophical justification to save the patient:

> Situations in life often permit no delay; and when we cannot determine the method which is certainly best, we must follow the one which is probably the best. . . . If the method selected is not indeed a good one, at least the reasons for selecting it are excellent.

Even though heuristics are nonanalytic, often false, and sometimes contradictory, they are properly used to solve problems that are so complex and poorly understood that conventional analytical techniques would be either inadequate or too time-consuming for use. The final vindication of the heuristic, irrational approach to solving problems over the rational, scientific approach would occur if the last survivor of a nuclear holocaust that had completely obliterated the entire human species were finally moved to announce with her dying breath, "At last I have a statistically significant sample with a $p < 0.01$ and can predict human behavior and avoid war." This ability to solve unsolvable problems or to reduce the search time for a satisfactory solution is the third characteristic by which a heuristic may be recognized.

The final signature of a heuristic is that its acceptance or validity is based on the pragmatic standard *it works or is useful in a specific context* instead of on the scientific standard *it is true or is consistent with an assumed, absolute reality.* For a scientific law the context or standard of acceptance remains valid but the law itself may change or become obsolete; for a heuristic the context or standard of acceptance may change or become obsolete but the heuristic—as heuristic—remains valid. Figure 9 will help make this distinction.

Science is based on conflict, criticism, or critical thought, on what has been called the Greek way of thinking. A new scientific theory, say B', replaces an old one, B, after a series of confrontations in which it is able to show that, as an approximation to reality, it is either broader in scope or simpler in form. If two scientific theories, B and B', predict different answers to a question posed by Nature, at least one of them must be wrong. In every scientific conflict there must be a winner. The victor is declared the best representative of "the way things really are" and the vanquished discarded as an interesting, but no longer valid, scientific relic. Ironically, the loser is often demoted to the rank of a heuristic and still used in cases of expediency. Thus, Einstein's theory replaced Newton's as scientific dogma and Newton's law of gravitation is used, in the jargon of the engineer, when a *quick and dirty* answer is needed. The traditional scientist assumes that the set, U, exists; that it does not change in time; and that it is eternal. It is given a special name, *reality*. Only the set of currently accepted scientific laws changes in time.

On the other hand, the absolute value of a heuristic is not established by conflict, but depends upon its usefulness in a specific context. If this context changes, the heuristic may become uninteresting and disappear from view, awaiting, perhaps, a change of fortune in the future. Unlike a scientific theory, a heuristic never dies; it just fades from use. A different interpretation of Figure 9 is therefore more appropriate in the case of a heuristic.

For the engineer, the set U represents all problems that are of interest to him at a given moment instead of all problems that are ultimately answerable. As a result, it is not a constant but varies in time. The engineer's set U ebbs as the obsolescence of the buggy has left the heuristics for buggy whip design high and dry on the shelf in the blacksmith's workshop, and it flows as renewed interest in self-sufficiency has sent young people in search of the wisdom of the pioneers. One heuristic does not replace another by confrontation, but by doing a better job in a given context. Both the engineer and Michelangelo "criticize by creation, not by finding fault."

The dependency on immediate context instead of absolute truth as a standard of validity is the final hallmark of a heuristic. It along with the other three signatures given are not the only important differences between the scientific law and the heuristic, but they are sufficient, I think, to indicate a clear distinction between the two.

This difference is summarized by viewing the campaign to conquer all knowledge, that is, the total area of U in Figure 9, from the perspective of the scientist and engineer with the former using scientific theories and the latter using heuristics as his troops. That the total crosshatched area advances relentlessly and essentially monotonically is an assumption of all scientists. That each individual crosshatched area will expand to eventually coalesce, covering the entire constant area U, is a further article of faith of the most radical among them. The scientist wages a campaign for these causes and coordinates the assault on the uncontrolled terrain. Strategically, she orders allegiance to the laws of thought (such as the laws of identity, contradiction, and the excluded middle); and tactically, to the laws of science (such as the requirement that a new theory be simpler or more universal than the one it replaces). Compared with this organized campaign, the engineer is engaged in guerrilla warfare. He sees no constant overall territory to be conquered and can be recognized by his ragtag troops. Instead of an army of laws and principles that are assumed to guarantee a victory, he has one of hints and rules of thumb that do not. Instead of troops that attack as a well-coordinated unit, his are often in disarray and squabbling among themselves. Instead of discharging an ineffective unit when a more effective one has been found, the engineer's soldiers follow the fortune of the immediate battle. But despite this disadvantage, while the scientist is trying to organize a professional search-and-destroy mission to conquer the terrain of all knowledge, the engineer has often already sought and destroyed. It remains to be seen which method proves the most powerful and universal.

Before we leave the set of all knowledge and the scientist's strategy to conquer it, two additional observations must be made. First, I feel compelled to express my curiosity about the epistemological status of the laws of thought and the laws of science in the previous paragraph. They seem so necessary to get the scientist's program off the ground. Is there any chance that they might just be heuristics themselves? Second, since no scientist would hold that she knows everything at the present moment, there must exist at least one fact about Nature that she does not now know in the unconquered area in Figure 9. Wouldn't it be the supreme irony if this fact was that knowledge does not or cannot exist? Indeed, there is a group of philosophers, the skeptics, who feel this fact does exist. The impact of such a lethal fact to the scientist's program would hardly faze the engineer, however. He would continue to build your homes, automobiles, and rockets and to develop the transportation systems to deliver the medicine and food you need to remain alive.

Synonyms for the Heuristic

Most engineers have never consciously thought about the formal concept of the heuristic, but all engineers recognize the need for a word to fit the four characteristics given. They frequently use the synonyms *rule of thumb, intuition, technique, hint, rule of craft, guiding thread, engineering judgment, working basis, guiding principles*, or, if

- in France, *le pif* (the nose),
- in Germany, *Faustregel* (the fist),
- in Japan,

   目
   の
   子
   勘
   定 * (measuring with the eye), or
- in Russia, *На пальчах*[†] (by the fingers),

to describe this plausible, if potentially fallible, experience-based strategy for solving problems. Each of these words captures the feeling of doubt that is characteristic of the heuristic. Each country just differs on which part of the anatomy is the most appropriate to use.

---

*Menoko Kanjo.
[†]Na paltsakh.

Examples of Engineering Heuristics

"Nothing of any value can be said on method except through example," or so counsels the eminent philosopher Bertrand Russell. Cowered by such an authoritative—well—heuristic, we conclude our present considerations with four concrete, well-documented examples of engineering heuristics.

Since we have seen that an engineering problem is defined by its resources, an appropriate place to begin is with a selection of the many heuristics the engineer uses to allocate and manage the resources available to him. For convenience, we will phrase each example in terms of an admonition. We will consider in some detail the following heuristics: *At the appropriate point in a project, freeze the design; Allocate resources as long as the cost of not knowing exceeds the cost of finding out; Allocate sufficient resources to the weak link;* and *Solve problems by successive approximation.*

I cannot certify from personal experience the observation of the English engineer Gordon Glegg, writing in *The Science of Design*:

> Rightly or wrongly, the U.S.A. has the reputation of being able to develop a new invention much more readily than we do in this country. If this is true, it may well be that one of the reasons for it is that the Americans usually veto any improvement in design after construction has begun. Leave it alone and alter the design in the next machine or the next batch; don't tinker with this one is their policy. And it is a highly realistic one.

If this statement accurately describes engineering design in England, it expresses a significant difference in the engineering practice of two countries with respect to the engineering

HEURISTIC: At some point in the project, freeze the design

because this heuristic is quite commonly used by the American engineer. Occasionally, it is explicitly expressed as in the book on nuclear reactor theory *Fast Reactors*, by Palmer and Plat, which reminds the reader that the set of numbers it gives was fixed by the size of the computer code when "the design was *frozen.*" This rule of thumb recognizes that a point is often reached in design where the character of a project, and hence the appropriate allocation of resources, changes from seeking alternative solutions to perfecting a chosen solution. As might be expected, this point is located heuristically by a trade-off between the relative risk and benefit of seeking yet another alternative. After this point is reached, a major design change runs an unacceptable risk of introducing a fatal flaw because insufficient resources remain to evaluate all of its ramifications. Once a design has been frozen (as a good rule of thumb, about 75 percent of the way into the project), the members of the design team take the general attitude, "Let's go with it."

The second item on the list, the engineering

HEURISTIC: Allocate resources as long as the cost of not knowing exceeds the cost of finding out,

appears frequently in the literature in a variety of forms. One author suggests "that a project be continued when confidence is high enough to permit further allocation of resources for the next phase or should be discontinued when confidence is relatively low"; another asks does "what has been learned about the project to date or the current prospects of yielding a satisfactory answer justify continuing to invest additional resources"; and a third author prefers the question, "Does what we know now warrant continuing?" Each of these formulations is essentially the same and simply acknowledges the trade-off between knowing and not knowing. In each case, two conflicting options carry an associated cost, and the engineer must decide, heuristically, of course, which cost is lower.

An interesting ramification of this rule of thumb condones the engineer's refusal to explore preposterous design alternatives. A subtle distinction exists between justifying consideration of an alternative view on the basis of heuristics, as is typical of the engineer, and on the basis of truth, as is typical of the scientist. Since truth is generally held to be an absolute good (and what scientist would not prefer a theory closer to the truth than one that was not?), in principle science must grant all points of view an audience no matter how bizarre. Despite the initial strangeness of the theory of relativity, it would have been bad scientific taste if it had been rejected before it had done battle against other theories to prove its mettle as a closer approximation to truth. This same criterion is not necessarily used to reject an alternative engineering design concept. Some years ago, a crackpot device called the Dean Machine was proposed to solve the problems of air travel. This device was alleged to be able to hover and fly for an indefinite period of time without any outside source of energy or any interaction with a known field. On the face of it, the Dean Machine violates essentially every known law of science beginning with Newton and ending with the second law of thermodynamics. No working model was provided, and the few available schematic diagrams showed a complex arrangement of eccentric cams that would have required many hours to analyze. Brandishing such taunts as, "They laughed at Copernicus" and "They laughed at Galileo," proponents of the Dean Machine accused the engineering fraternity of being closed-minded, cliquish, and afraid of unconventional ideas because it would not sponsor research to prove whether or not the Dean Machine would work. Although science has no rule of procedure to dismiss a theory before the battle for truth is joined, engineering does. Since the analysis of the Dean Machine would require a large amount of the available resources, since it represents a very large change in the current engineering knowledge, and since it could hardly be called best engineering practice, even after only a cursory glance, the engineer can justify rejecting further study because "what he now knows does not warrant continuing."

Of course, this heuristic, as with all others, does not guarantee that the absolute best decision has been made. The Dean Machine just might have worked. But considering the large penalty associated with resources squandered on a wild goose chase, if the available resources restrict the engineer to the well-traveled path, he need not regret the opportunity missed on the road not taken.

The next item on the list of example engineering heuristics, the

HEURISTIC: Allocate sufficient resources to the weak link,

ostensibly refers to the English aphorism, "A chain is as strong as its weakest link." By extension, it implies that if a stronger chain is desired, the correct strategy is to strengthen the weak link. For purposes of illustration, limit the project to put a man on the moon to the major subsystems: telemetry, life-support, guidance, launch vehicle, and return vehicle. If we further assume that the safe return of the man depends on the successful operation of each of these subsystems, the importance of the weak link is obvious. During the design to put a man on the moon, resources must be allocated to the least assured system to increase the probability of mission success.

All engineering designs have this same problem. With the possible exception of the famous one-hoss shay described by Oliver Wendell Holmes in the poem "Deacon's Masterpiece or the One-Hoss Shay," which is reputed to have had no weak link, every engineering project has a limiting element in its design. Good engineering practice requires that sufficient resources be allocated to this element. The overall design will be no better and, because of overdesign of the lesser important parts of the final project, may actually be worse, if additional resources are allocated to the less critical components. If we are to believe Mr. Holmes, perfect engineering design is achieved only when the shay has no weak link, and all parts of a design collapse at the same time into a heap of dust after "one hundred years and a day."

The same concept exists in other disciplines, chemistry for one, where the weak link is now called the limiting reagent. Since one atom of sodium combines with one atom of chlorine to produce one molecule of common table salt, the final amount of salt in this chemical reaction depends upon which of the original elements, the limiting reagent, is in short supply. In this, as in all problems of chemical stoichiometry, if you want more product, increase the limiting reagent; in effect, attack the weak link.

This heuristic is so important that in certain limited cases the engineer has produced a theoretical formulation for it such as in the algorithms for scheduling and coordinating the many individual tasks needed to complete most large engineering projects. Construction of a building, airplane, or bridge requires that blueprints be drawn up, the site be prepared, materials be procured and verified, personnel be hired, and so on. Some of these tasks can be performed in parallel; others must await the completion of an earlier task before they can be started. The Critical Path Method and the related Performance Evaluation and Review Techniques (often called by their engineering acronyms CPM and PERT analysis) are mathematical strategies for scheduling individual tasks and finding the critical sequence of them (or as we would say, the weak link) for completion of a project. The overall time to construct a bridge is not shortened if additional resources are allocated to finish tasks not found on the critical path. Theory predicts that if you want to decrease the time to complete the project, attack this weak link.

Two concerns may have arisen as we consider this simple heuristic that aids the engineer in allocating his resources. First, you may feel that the admonition to allocate resources to the weak link just makes common sense and that it is always the best policy. There is no need to come up with a special name for it by calling it a heuristic because everyone uses it—or so the first concern goes. Unfortunately this is not the case. Everyone does not use it. For example, professors observe that students in scholastic difficulty do not allocate study time (a resource) to study the course that will do the most to improve their overall grade point average. Instead students typically choose to study the courses they prefer, the ones they find the easiest, and the ones in which they are doing well. Likewise, a homeowner trying to improve the appearance of her home does not look for the change that will cause the greatest improvement, but rather she seeks out the easiest, the most convenient, the least expensive, or the best understood one. Common sense or not, application of this heuristic is far from common.

The second concern with this heuristic is that you may feel that irrespective of whether or not humans always use it, they should. That is, you may feel that it is a truism. Unfortunately this supposed truism is far from always true. Allocation to the weak link is a heuristic because it is not always appropriate. A quotation from a well-known book on engineering will refute this view. Let me quote from *Reliability Technology* by Green and Bourne:

> The development of the V1 [missile in Germany during the Second World War] started in 1942 and the original concept for its reliability *was that a chain cannot be made stronger than its weakest link.* Although efforts were directed along this line, there was 100% failure of all missions in spite of all the resources available under wartime conditions. Eventually, it was realized that a large number of fairly strong "links" can be more unreliable than a single "weak link" if reliance is being placed on them all. Work based on this idea gave rise to a great improvement in the reliability of the V1. It is interesting to note that the reliability ultimately achieved was that 60% of all flights were successful and a similar figure also applied to the V2 missile. (emphasis mine)

The final example of a heuristic that engineers use to allocate project resources we will give is the

HEURISTIC: Solve problems by successive approximations.

It is of particular interest because it also shows the wide applicability of engineering heuristics to other areas in which problems are solved. A previous heuristic counsels us to attack the weak link, but how are we to find this weak link? The engineering answer is to begin with an overall view of a problem and then to perfect it by successively using a series of more detailed approximations. Reconsidering the example given earlier, the engineer would begin with a first cut at the telemetry, life-support, guidance, launch vehicle, and return vehicle systems and then design each in more and more detail. I remember one of my engineering design professors admonishing our class to first find the baseball park, then locate the baseball diamond, and finally look for home plate to express this heuristic of looking at the big picture first and perfecting the design in smaller and smaller stages.

This strategy does not appear to be commonly used by humans. The clear tendency is to begin at the beginning of a problem, to march through the project, and, when you get to the end, to stop. In fact, the natural tendency to ignore this heuristic is easy to demonstrate in the laboratory. In sophomore electrical engineering courses, for instance, students are often required to balance a strange device called a Wheatstone bridge. The exact nature of this device does not concern us here, but basically the student is asked to locate the point along a wire at which the bridge gives a zero reading. Most students begin at one end of the wire and move timidly along it in small steps until they reach the correct point. This can often take a very long time. A better procedure is to divide the wire in half and to locate in which half the correct value is to be found and then further subdivide the correct half and so forth.

The heuristic that suggests we solve problems by successive approximations is valuable for the artist and regularly taught in art classes. A painter does not usually complete one small area of her painting before moving on to the next, but instead develops the entire canvas at the same time. If her picture is to be a deer in a sun-drenched glen, she does not paint the deer down to the last antler on Monday, the glen on Tuesday, and worry about drenching the whole in sunlight on Wednesday. An extreme example of this process of metamorphosis is seen in the work of the painter Picasso. Often in his paintings an ear or eye of one figure seems to migrate to become a part of another one throughout the process of painting. From the initial sketch through the final masterpiece, a better and better representation of the total scene emerges by successive approximations.

In an article entitled "Guernica and 'Guernica'" in *Artforum*, Phyllis Tuchman quotes Picasso on this process:

> It would be very interesting to preserve photographically, not the stages, but the metamorphoses of a picture. Possibly one might then discover the path followed by the brain in materializing a dream. But there is one very odd thing— to notice that basically a picture doesn't change, that the first "vision" remains almost intact, in spite of appearance.

Fortunately we have been left with visual evidence that Picasso implemented the heuristic that an artist should complete a project by successive approximations. Three pictures from a large series of photographs of the stages of Picasso's famous painting *Guernica* are given in Figures 10, 11, and 12.

How like an engineering project does this seem? I, along with my engineering colleagues, nodded in agreement with Picasso with only minor changes as Neil Armstrong stepped out on the moon, answering President Kennedy's initial call to put a man there by the end of the decade.

Figure 10 Study for *Guernica. (Source: © Estate of Pablo Picasso / Artists Rights Society (ARS), New York)*

Figure 11 Study for *Guernica. (Source: © Estate of Pablo Picasso / Artists Rights Society (ARS), New York)*

The same can be said of all art. The novelist writing a book, the sculptor chiseling a statue, the musician composing a concerto, the engineer designing a bridge, and the theater director giving her actors the first crude stage blocking all allocate their resources so that improvement in their product is by successive approximations.

The four heuristics—At the appropriate point, freeze the design; Allocate resources as long as the cost of not knowing exceeds the cost of finding out; Attack the weak link; and Solve problems by successive approximations—are not the only resource allocation heuristics in the armory of the engineer. If the time we had together were longer, we could stop to consider many other interesting rules of thumb in this category. The heuristics given, however, are excellent examples of this important class and worthy of special study. They are included here for two reasons: First, they give concrete examples of engineering heuristics, and second, they hint at the universality of some heuristics in all methods of problem solution.

No claim is being made that only engineers and practitioners of the traditional liberal arts use heuristics to solve problems. Politicians, journalists, and

Figure 12 *Guernica,* by Picasso. *(Source: © Estate of Pablo Picasso / Artists Rights Society (ARS), New York)*

doctors also use them. Politicians, for example, have their own unique rules of thumb for solving problems—the principal of which seems to be getting re-elected. When American politicians say *Democrats cannot win without the labor and minority vote; Republicans cannot win without the vote of business; Define your-self before others do;* and *Don't step on your own news story,* they are passing along political rules of thumb. Each of these heuristics provides a plausible aid or di-rection in the solution of a problem, but each is in the final analysis unjustified, incapable of justification, and potentially fallible.

Political heuristics are often useful in averting disaster for a party, such as when the Democratic party once gave a senator political cover, a fig leaf as it is called, so he could vote against his conscience and cast the deciding vote for the president's deficit reduction package. Of course, being rules of thumb, these pithy sayings are fallible, as another congresswoman learned to her chagrin. The political rule of thumb *The party has a longer memory than the electorate* proved to be her undoing when she ran for election on one promise, but voted differently as her party dictated once elected. Her constituents did not let her forget it and voted her out of office.

Likewise journalists remind us that *When it bleeds, it leads; If you want it pub-lished, it is advertising, if we want it published, it is news;* and *Theoretically journal-ists cannot be objective, but they are fair.* In an article entitled, "Washington's 'Bermuda Triangle of News'" in the *Washington Post,* journalist R. Jeffrey Smith quotes an advisor to presidents, reminding us of what can only be considered a heuristic: "There is a tradition that goes back to the Dark Ages, which says that the best time to release bad news is Friday after 4 p.m. Some of us have called that [time period] the Bermuda Triangle of news because anything that goes in never comes out." Similarly, doctors once bled patients with leeches, then this practice was discontinued, and now I understand that the limited use of leeches is being reconsidered for other medical purposes. In fact, medicine as a whole is nothing but a set of heuristics subjected to change with era, culture, and whim.

Finally, the attentive philosopher will no doubt anticipate that, in good time, I will also assert that she uses heuristics. I have already expressed misgivings about the epistemological status of her concepts: *classification, change, definition,* and *set.* Should we be equally suspicious about the concepts: *logic, truth, causal-ity,* and *perception*?

This completes the consideration of the technical word *heuristic* that we needed for the definition of engineering method until Part III where an exten-sive list of examples is given. We have analyzed the important concept of the heuristic by analogy with the hints and suggestions given to learn to play chess, by definition, by looking at four signatures that distinguish it from a scientific law, by reviewing a list of its synonyms, and through several examples. It seems undeniable that at times engineers use heuristics. All engineers seem to agree on this fact, although they may use different terms for the concept.

STATE OF THE ART

Instead of a single heuristic used in isolation, a group of heuristics is usually re-quired to solve most engineering design problems. This introduces the second important technical term, *state of the art*. If you have been in the presence of an engineer for any length of time, you will have heard him slip this term into the conversation. He will proudly announce that his stereo has a state-of-the-art speaker system or that the state of the art of computer design in his home country is more advanced than elsewhere. Since this concept is fundamental to the art of engineering, attention now shifts to the definition, evolution, and transmission of the state of the art, along with examples of its use.

Definition

State of the art, as a noun or as an adjective, always refers to a set of heuristics. Since many different sets of heuristics are possible, many different state of the arts exist and to avoid confusion each should carry a label to indicate which one is under discussion. Each set, like milk in the grocery store, should also be dated with a time stamp to indicate when it is safe for use. Too often in the past, the neglect of the label with its time stamp has caused mischief. With these two exceptions, no restrictions apply to a set of heuristics for it to qualify as a state of the art.

In the simplest, but less familiar, sense, state of the art is used as a noun referring to the set of heuristics used by a specific engineer to solve a specific problem at a specific time. The implicit label reminds us of the engineer and the problem, and the time stamp tells us when the design was made. For example, if an engineer wants to design a bookcase for an American student, he calls on the rules of thumb for the size and weight of the typical American textbook, engineering experience for the choice of construction materials and their physical properties, standard anatomical assumptions about how high the average American student can reach, and so forth. The state of the art used by this engineer to solve this problem at this moment is the set of these heuristics. If the same engineer was given a different problem, such as the design of a bookcase for a French student, he would use a different group of heuristics and hence a different state of the art. Bookcase design is not the same in the United States as it is in France. In the former, the title of a thin book is always printed on the spine with the bottoms of the letters near the back of the book in such a way that the title reads correctly—from below—when the book is lying on its back. In the latter, the title is often reversed so that it reads correctly—this time from above—when the book is in the same position. Other differences between the two countries are better known. American books are measured in inches; French books, in centimeters. The average French student is smaller than her American counterpart and prefers different construction materials, and so on. As a result, the set of heuristics that are appropriate for use in America and France differ. Since a book in China reads from what I would consider to be the back of the book to the front, still a third state of the art would apply there.

Returning from our world tour, let us consider two engineers who have been given the same problem of designing a bookcase for an American student. Each will produce a different design. Since a product is necessarily consistent with the specific set of heuristics used to produce it and since no two engineers have exactly the same education and past experience, each will have access to similar, but distinctly different, sets of heuristics and hence will create a different solution to the same problem. State of the art as a noun refers to the actual set of heuristics used by each of these engineers.

In a complicated, but more conventional, sense, state of the art also refers to the set of heuristics judged to represent *best engineering practice*. When a person says that her stereo has a state-of-the-art speaker system or that she has a state-of-the-art bookcase, she does not just mean that they are consistent with the heuristics used in their design. That much she takes for granted. Instead she is expressing the stronger view that a representative panel of qualified experts would judge her speaker system or her bookcase to be consistent with the best set of heuristics available. Once again, state of the art refers to an identifiable set of heuristics.

Because the design of a bookcase requires only simple, unrelated heuristics, it misrepresents the complexity of the engineering state of the art needed to solve an actual problem. More typically, the state of the art is an interrelated network of heuristics that control, that inhibit, and that reinforce each other. As earlier noted several times, several rules of thumb are available to evaluate the number of Ping-Pong balls in a room, and, therefore, an additional heuristic is needed to aid in selecting the appropriate rule of thumb in each situation. This heuristic does not contribute directly to estimating the number of Ping-Pong balls, but it controls the choice between rival strategies. Or, to use a more relevant engineering example: One of the physical properties of a large organic molecule called the *enthalpy* may be determined either in the laboratory by experiment or by estimating the number of carbon and hydrogen atoms that it contains and then applying a known formula. In practice the engineer sometimes uses one method to evaluate the enthalpy; sometimes, the other. Obviously he has another heuristic—perhaps something like, "Go to the laboratory if you need 10 percent accuracy and have $5,000"—to guide his selection between the two. Bookcase design, the evaluation of the number of Ping-Pong balls in a room, and the determination of the enthalpy of a large organic molecule are such simple examples that they hardly suggest the complexity of a state of the art. It is to your imagination that I must finally turn to visualize how much more complicated the state of the art used in the design of an airplane must be as the heuristics of heat transfer, economics, strength of materials, and so forth influence, control, and modify each other. The interaction and interrelation of the myriad heuristics in actual engineering practice are almost beyond comprehension.

Whether it is the set of heuristics that were actually used in a specific design problem or the set that someone feels would be the best one for use, state of the art always refers to a collection of heuristics—most often a very complicated collection of heuristics at that. Its imaginary label must let us know which engineer, which problem, and which set are under consideration.

Evolution

The state of the art is a function of time. It changes as new heuristics become useful and are added to it and as old ones become obsolete and are deleted. The bookcase designed for a Benedictine monk today is different from the one designed for Saint Benedict in 530 A.D. When we discussed the design of a bookcase for students, I did not emphasize the time stamp that must be associated with every state of the art. Now is the time to correct this omission and consider the evolution of a set of heuristics in detail, beginning with a well-documented example.

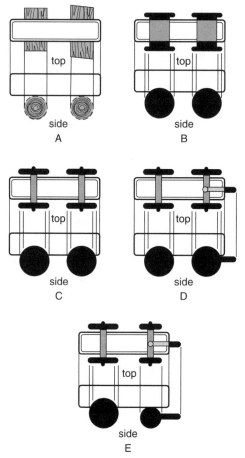

Figure 13 Evolution of Cart

In overall outline, scholars feel that the evolution of the present-day cart took place in a series of stages as shown in Figure 13. Since the wheel probably evolved from the logs used as rollers, it is assumed that the earliest carts had wheels rigidly mounted to their axles, wheels and axle rotating as a unit. This assembly has the disadvantage that one wheel must skid in going around a corner. As an improvement, the second stage was probably a cart with the axle permanently fixed to the body and with each wheel rotating independently. In this design neither the wheels nor the axle were capable of pivoting. It has the disadvantage that the cart cannot go around corners unless manhandled into each new position. This same difficulty still bothers mothers who own the old-fashioned baby stroller with fixed wheels. After twenty or thirty centuries, the engineer learned how to correct this problem by allowing the front axle to pivot on a kingbolt as stage three in the evolution of cart design. Since the front and back wheels were large and the same size, this cart could not turn sharply without the front wheels scraping against its body. In the final stage, the front wheels were reduced in size and allowed to pass under the bed of the wagon as the

MERCEDES-BENZ
REAR SUSPENSION DESIGNS

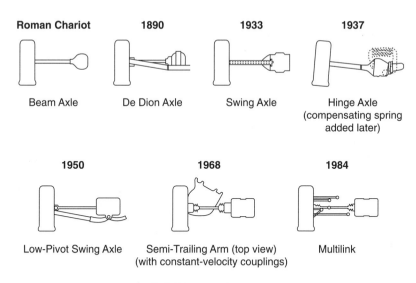

Roman Chariot

Beam Axle

1890

De Dion Axle

1933

Swing Axle

1937

Hinge Axle
(compensating spring
added later)

1950

Low-Pivot Swing Axle

1968

Semi-Trailing Arm (top view)
(with constant-velocity couplings)

1984

Multilink

Ref. *Mercedes XI, 1983*

Figure 14 Rear Suspension

front axle pivoted. This process of evolution has continued into the present day, of course, as the cart has become the automobile. But as we are short on time and this modern state of the art is more complicated than you might think, you will have to ask your local racing enthusiast or mechanical engineer what rack-and-pinion steering is and how it works. Can we doubt that the heuristics of cart design have changed over time?

Now please look carefully at pages 2 and 3 of the Color Insert (between pages 148 and 149) depicting the evolution of the Mercedes automobile. In the one hundred years shown, what is the predominate thing that has changed? The state of the art, of course.

Not only the overall design, but each of its individual components, down to the smallest detail, also evolves over time. The change of heuristics on this more detailed level is illustrated in Figure 14, of the rear suspension system of the Mercedes-Benz, from the earliest Roman chariot to 1984. Once again, the basic question that needs to be asked is this: Exactly what changed over this period of time? The only reasonable answer: the state of the art, the collection or set of heuristics used.

Physical devices are not the only things that evolve over time. Manufacturing techniques, strategies, and procedures have also changed. An excellent example of this is the change of heuristics in the manufacturing process of automobiles. In the beginning automobiles were made individually following the lead of the German, Karl Benz, who first brought them to the stage of commercial feasibility in 1886. It was for Henry Ford to invent the assembly line in 1912 and to introduce automation about 1949. Later in 1975 the Volvo factory in Kalmar decided that workers were being dehumanized by the inhuman assembly line.

Charlie Chaplin parodied this dehumanization in a classic sequence in the movie *Modern Times*. Kalmar introduced a procedure in which a team of workers followed a car down the assembly line completing all parts of the assembly. The procedure used at Kalmar was mandated by law at the Volvo factory at Udevalla in 1987. In more recent times, the Japanese have improved on the procedure of automotive manufacture by introducing two radically new heuristics: *total quality management*, which allows workers more responsibility over the organization and performance of their work, and *just in time* strategies, which streamlined the delivery of parts used in production. The basic point is that not only the product but also the strategies for producing the product change over time as new heuristics are added and removed from the state of the art.

From the twilight of time, early craftspersons used this same process of evolution, often producing near optimum results, as seen in the common bow and arrow. Theoretically, a bow will develop maximum power for its size when it acts as, what an engineer would call, a uniformly stressed beam. Recent analysis of the classical bow with its unusual shape has confirmed that it is indeed equivalent to such a beam. The development of early mechanical devices and weapons continued through centuries of trial and error as lucky accidents modified the state of the art until it contained many of the heuristics the modern engineer uses today. At every step along the way, each cart, each stirrup, each bow, and each automobile has stood as a monument to some engineer's idea of best, a monument valid as a standard of best only on its dedication day.

I wonder if the heuristics used by the politician, the journalist, and the doctor, not to mention the philosopher, have evolved in time?

Transmission

Down through the ages the state of the art has been preserved, modified, and transmitted from one individual to another in a variety of ways. The earliest method was surely a simple apprentice system in which craftspersons carefully taught rules of thumb for firing clay and chipping flint to their assistants who would someday replace them. With hieroglyphics, cave paintings, and, later, books, the process became more efficient and was no longer dependent on a direct link between teacher and taught. Finally, in more recent times, trade schools and colleges began to specialize in teaching engineers. Despite the importance of apprentice, book, and school in preserving, modifying, and transmitting accumulated engineering knowledge, they need not detain us longer because of their familiarity.

An additional method is less well known and worth a minute's discussion. If an engineer is asked to design a cart today, he would use an articulated front axle and wheels that were small enough to pass under the bed of the cart. He would not do so because he is familiar with the evolution of cart design, but because the carts he actually sees around him today are constructed in this way. All traces of the state of the art that dictated axle and wheels turning as a unit have disappeared. The engineer does not need to know the history of cart design; the cart itself preserves a large portion of the state of the art that was used in its construction. In other words, modern design does not recapitulate the history of ancient design.

Since the heuristics of tomorrow are embodied in the concrete objects of today, the engineer is unusually sensitive to the physical world around him and uses this knowledge in his design. I once asked an architectural engineer* to estimate the number of Ping-Pong balls that could be put into the room in which we had been sitting. He quickly gave an answer by remembering that the room had three concrete columns along the side wall and two along the front. Since he also knew the rule of standard column spacing, he could calculate its size quite accurately. Because I was not trained as an architectural engineer, I didn't even remember that the room contained columns. This awareness of the present world translates directly into the heuristics used to create a new one. If a proposed room, airplane, reactor, or bridge deviates too far from what he has come to expect, the engineer will question, recalculate, and challenge. Although he may be unaware that he is wearing glasses, the engineer judges, creates, and sees his world through lenses ground to the prescription of his own state of the art.

Almost without knowing it, an engineering student establishes this sensitivity to the existent world as he solves home problems in which the answers are taken to be representative of good design, visits operating plants, looks around himself at the world created by older engineers, and completes his initial assignments after graduation. That is, the young engineer develops his own set of heuristics as he lives and grows in the state of the art of the modern engineer. I remember well the first time, as a young engineer, I saw the nuclear reactor at Marcoule, France. I had solved many nuclear reactor design problems in college and visited many nuclear installations in the United States. All of the reactors I had seen were relatively small. I was unprepared for the French reactor. It—was—huge. I wish I could capture for you the acute disorientation I felt as we looked down on one side of this four-story high nuclear reactor. It developed, of course, that the Marcoule reactor was gas-cooled and had a mammoth prestressed concrete outer shell of a type not used in the United States. Little wonder I felt strange, but as a result of this experience my state of the art for reactor design was permanently changed.

Sometimes using the old world as a template for the new goes astray and produces curious results. When engineers were given the task of creating the best playing surface for athletes, their solution consisted of a rug with tiny blades of green plastic, Astroturf as it was called. How could something that looks like grass, Nature's state of the art for trapping sun energy through photosynthesis, be justified as necessarily the best surface for playing football? The world as it is supplements on-the-job training, the technical literature, and colleges of engineering in providing the continuity and change in the state of the art used by the engineer. An imaginary time stamp on all engineering products is essential to remind us where we are in this evolutionary process.

To review: The state of the art is a specific set of heuristics designated by a label and time stamp. It changes in time and is passed from engineer to engineer either directly, by the technical literature, or in a completed design. Typically it includes heuristics that aid directly in design, those that guide the use of other heuristics, and, as we will see later, those that determine an engineer's attitude or behavior in solving problems. The state of the art is the context, tradition, or environment (in its broadest sense) in which a heuristic exists and based upon which a specific heuristic is selected for use. We might also characterize the state of the art of the engineer as his privileged point of view.

---

*Dr. David Fowler.

I can't but wonder if the world, culture, country, religion, and society into which each of us is born acts as a template and fashions our reality in a similar way.

The term *state of the art* is currently making its bid to become a full member of the English language, not only as applied to technology, but also in other situations where it is logical to speak of a set of heuristics that represent the current best idea of how something should be done. Its orthography has not, however, stabilized as either "state of the art," state-of-the-art, or state of the art. It is not surprising when we read in the newspaper that "a new telephone system is '*state-of-the-art*'"; slightly more so when we read that "the new [dollar] bills will almost certainly contain *state-of-the-art* security threads"; and definitely so that "the *state of the art* of winemaking changed rapidly, attended as it was by the best minds of the times." These are but a few isolated examples of many recent attempts to broaden the scope of the term and enter it into the conceptual framework of English as a term implying the current best way of doing something.

An Acronym for State of the Art

State of the art, no matter how it is written and used, is both a cumbersome and inelegant term. After only a few pages of definition, we have become tired of seeing it in print. Therefore, from now on I propose to replace it by its acronym *sota*. Coining a new word by using the first letters in an expression's constituent words is a familiar procedure to the engineer who speaks freely of *radar* (radio detecting and ranging) and the *lem* (lunar excursion module). In large engineering projects such as the manned landing on the moon, acronyms are so frequently used that often a project description appears to the nonengineer as written in a foreign language. At times an acronym even takes on a life of its own, and we forget the words used to create it. Few remember the original definitions of *laser*, *scuba*, *zip code*, and, perhaps fortunately, *snafu*. I wish the same fate for this new acronym *sota*.

The philosopher, however, quickly grows impatient with the engineer's constant use of these shortened forms, although ironically she has developed her own procedure for identifying and emphasizing the important concepts essential to her work. Her usual strategy is to borrow words from a foreign language. This tendency to naturalize foreign words such as *organum*, *zeitgeist*, and *cogito* is equally obnoxious to most engineers. Both disciplines feel the need to set apart a word for technical use; each employs a different heuristic to accomplish this purpose. From now on *sota*, used both as adjective and noun, is to be taken as a technical term meaning an identifiable set of heuristics.

Example Uses of the sota

Without due consideration, the concept of an engineering state of the art as a collection of heuristics appears contrived and its acronym gimmicky. The frequency with which the word *sota* will appear in the remainder of this discussion and the relief we will feel at not seeing its expanded form each time will answer the second criticism. Three specific examples showing the effectiveness of the sota as a tool for bringing understanding to important aspects of the engineering world will answer the first. Various sets of heuristics will now be used:

1. To compare individual engineers
2. To establish a rule for judging the performance of an engineer
3. To define the relationship between the engineer and society

This last example will also suggest the importance of technological literacy for the nonengineer and liberal literacy for the engineer. Although these examples will occupy a reasonable amount of our time, they will dispel any feeling of artificiality in the notion of a sota and suggest important areas for future research. We will then, at last, be in a position to consider the principal rule for implementing the engineering method.

Comparison of Engineers

The individual engineer, in his role as engineer, is defined by the set of heuristics he uses in his work, including the heuristics he has learned in school, developed by experience, and gleaned from the physical world around him. When his sota changes, so does his proficiency as an engineer. The area inside of the closed boundary in Figure 15 will represent this set. Characteristic of all sotas, this one needs a label to differentiate it from others and a time stamp to indicate when it was evaluated. Using the symbol $sota|_{A;t}$ for the state of the art of Mr. A at time t will greatly simplify the discussion. The symbol representing my sota as an engineer when the first written version of this book appeared is, of course, $sota|_{Koen;1982}$, an improvement, I hope, over $sota|_{Koen;1965}$, when I first began learning to play chess and the concept of an engineering heuristic was born.

No two engineers are alike. The first example we will consider of the use of a sota will make this point. The sotas of three engineers, A, B, and N, are shown in Figure 16. They share those heuristics inside of the area indicated by the small rectangle where they overlap, but each also encloses an additional area to account for the unique background and experience of the engineer it represents. In general, if A, B, and N are all chemical engineers, the overlap of their sotas is larger than if A is a civil engineer, B is a chemical engineer, and N is a mechanical engineer. Most chemical engineers have read the same journals, attended the same conferences, and quite possibly used the same textbooks in school. Not surprisingly, they share many engineering heuristics. This rule of thumb may not always be true, of course. A chemical engineer who has specialized in hydraulics may have more in common with his counterpart in mechanical engineering who has also specialized in hydraulics than with his other

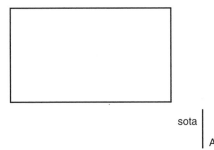

sota |
    A;t

Figure 15 sota of Individual Engineer

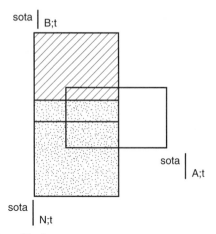

Figure 16 sotas of Three Engineers

colleagues in chemical engineering. If the sotas of all modern chemical engineers, instead of only the three shown here, were superimposed, the common overlap of all of these sets, technically called the intersection, would contain the heuristics required to define a person as a modern chemical engineer. Both society and the engineering profession have a vested interest in preserving the integrity of this area. The label *chemical engineer* must ensure a standard approach and minimum level of competency in solving chemical engineering problems. We will return to the intersection of all engineering sotas later in our discussion when we discuss the heuristics it must contain for a person to be properly called an engineer. For now, remember that the overlap of the appropriate sotas measures the similarity and dissimilarity of engineers. You may guess that I will consider a sota of the philosopher as sota|$_{phil.}$ later. How much overlap should we expect this sota to have with an engineer's sota?

Rule of Judgment

No one, I think, would argue that an engineer should not be held responsible for his work. The difficulty is knowing when he has done a satisfactory job. A discussion of the correct rule for judging the engineer will be a good second example of using an engineering sota and will emphasize, once again, the importance of attaching an imaginary label and a time stamp to each one. This example will also point out the difficulty of implementing the Rule of Judgment.

Theoretically, we would like to hold an engineer responsible for designing the product he ought to design. But *ought* implies ethics, and the study of ethics is in a mess. If *classification, change, definition,* and *set* have caused disarray among philosophers, *ethics* has caused even more. Is the state, the existing situation, the individual, a religion, or some absolute standard to be the final arbiter of the good and the ethical? The engineer seeking professional guidance quickly finds himself drowning in the sea of "ist." Should he believe the intuitionist, the empiricist, the rationalist, the hedonist, the instrumentalist, the situationalist, the pragmatist, or (if this is an acceptable word for one who endorses Ayer's emotivism) the emotivist? But, once again, I stray into the domain of the philoso-

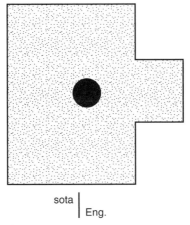

sota | Eng.

Figure 17 sota of All Engineers

pher. It seems to the engineer that everyone has an opinion, but that no one has a definitive answer. With all of this disagreement, little wonder the engineer fears to tread where the philosopher rushes in. Unfortunately the engineer cannot wait, for you want your bridge, your home, or your automobile tomorrow; and, therefore, until the specialists reach agreement, the engineer must rely on his own definition of "ought."

The outside border in Figure 16 indicates the set of all heuristics used by A, B, and N, at time, t, or $sota|_{A,B,N;t}$. If all engineers were included in this figure, it would delimit the sota of the engineering profession as a whole, or $sota|_{eng.\ prof.;t}$. The stippled area in Figure 17 reproduces this sota at a given time, t.

No engineer will have access to all of the heuristics known to engineering, but in principle some engineer somewhere has access to each heuristic represented in this figure. The black, solid circle represents the subset of heuristics needed to solve a specific problem. The combined wisdom of the engineering profession defines the best possible engineering solution to it. This overall sota represents best engineering practice and is the most reasonable practical standard against which to judge the individual engineer. It is a relative standard instead of an absolute one, and like all sotas it changes in time.

To my knowledge, no engineer is clairvoyant. Handicapped in this way, it would seem unreasonable to expect him to make a decision at one moment based on information that will only become available to him later. He can only make a decision based on the set of heuristics that bears the time stamp certifying its validity to him at the time the design must be made. With these considerations in mind,

> The fundamental *Rule of Judgment* in engineering is to evaluate an engineer or his engineering design against the sota that defines best practice at the time the design was made.

This rule is logical, defensible, and easy to state. Unfortunately, it is not universally applied through ignorance, inattention, and a genuine difficulty in extracting the sota that represents good engineering practice from the set of all engineering heuristics in a specific case. Each of these three reasons for not observing the Rule of Judgment is worthy of special attention.

For the layperson, the failure of an engineering product usually means that some engineer somewhere has done a poor job of design. This criticism is based on ignorance of the correct basis for judging the engineer and is indefensible for two reasons. First, an engineering design always incorporates a finite probability of failure. The engineer uses a complex network of heuristics to create the new in the area of uncertainty at the margin of solvable problems. Hence, some failures are inevitable. Had the ancient engineer remained huddled in the security of the certain, he would never have ventured forth to create the wheel or the bow. The engineer should not be criticized by looking only at a specific failure while ignoring the context or sota that represents best engineering practice upon which the decisions that led to that failure should have been based. Second, the world changes around the completed engineering product. A sota tractor on the date it was delivered is not necessarily appropriate for working steeper terraces and pulling heavier loads fifty years later, and it may fail or overturn or both. Often the correct basis for judging the performance of an engineer is not used because the public, including juries charged with deciding cases involving product liability and journalists reporting major technological failures, does not know what it is.

Engineers are also misjudged through inattention. Since the sota is a function of time, special attention is needed to ensure that the engineer is evaluated against the one valid at the time he made his design. Two examples will demonstrate how easy it is to forget this requirement, easy even for an engineer.

One of my French engineering colleagues, undoubtedly carried away with nationalistic zeal, surveyed the modern Charles de Gaulle Airport in Paris, France, and exclaimed that he had just had a miserable time getting through O'Hare Airport in Chicago. He then added that he could never understand why American engineers are not as good at designing airports as are the French. His mistake was wanting a clairvoyant engineer. Leaving aside the factor of scale— the American airport was, at the time, the busiest in the world, de Gaulle had only been in operation two weeks—two facts are beyond dispute: Each airport was consistent with the sota at the time it was built, and, given the comprehensive exchange of technical information at international meetings, the sota on which the French airport was based was surely a direct outgrowth of the earlier one used in Chicago. Even an engineer sometimes forgets the time stamp required on an engineering product.

The second example concerns the aphorism of the American frontier that a stream renews itself every ten miles. Essentially this means that a stream is a buffered ecosystem capable of neutralizing the effects of an incursion on it within a short distance. Let us assume that an enterprising pioneer built a paper mill on a stream and discharged his waste into it. According to this aphorism, no damage was done. Now let us add that over the subsequent decades additional mills were constructed until the buffering capacity of the stream was exceeded and the ecosystem collapsed. Your job is to fix blame. If you argue that the later engineers were wrong to use a heuristic that was no longer valid, I agree. Presumably the original heuristic is applicable only to a virgin stream. If you argue that the original plant should be modified to make it consistent with later practice, a process the engineer calls retro- or back-fitting, I might also agree, although I am curious whether you would require the pioneer, later plant owners, or society to pay for this often expensive process. But if you criticize the original decision to build a plant on the basis of today's set of heuristics, I most certainly do not concur. As others have said, we must judge the past by its own rule book, not by ours.

These two examples show how easy it is to forget the factor of time in engineering design through inattention. But now let us consider a more troublesome reason why the engineer is often not judged on the basis of good engineering practice at the time of his design. The problem is agreeing on what sota is to be taken as representative of good engineering practice. All engineers cannot be asked for their opinions; that is, $sota|_{eng.prof.}$ cannot be used as a standard. The only recourse is to rely on a "panel of qualified experts" to give its opinion. But how is such a panel to be constituted? Is membership to be based on age, reputation, or experience? In determining the set of heuristics to represent a sota chemical plant in America, should foreign engineers be consulted? And finally, when best engineering practice is used as a basis for how safe is safe enough for a nuclear reactor, should members of the Sierra Club be included? No absolute answers can be given. But the engineer has never been put off by a lack of information and is willing to choose the needed experts—heuristically. Like any other sota, the set of heuristics he uses to choose his panel will vary in time and must represent best engineering practice at the time he constitutes it.

Agreement about which sota should represent best engineering design at the present moment is hard enough, but agreement about the set of heuristics that was appropriate fifty years ago is even harder. Many of the designers of engineering projects still in use today are no longer alive. Was the steel in the Eiffel Tower consistent with the best engineering practice of its day? With no official contemporary record to document good engineering judgment, history easily erases our memory of what was the appropriate sota for use in the past. Given the recent rash of product liability claims against the engineer, what is now needed is an archival $sota|_{best\ eng.judge.}$ to allow effective implementation of the Rule of Judgment.

Engineer and Society

The relationship between the engineer and society is the last, and most extensive, example we will consider of the use of various sotas. It is also one of the most important.

All heuristics are not engineering heuristics, and all sotas are not engineering sotas. As has been observed by other authors, aphorisms, which have all of the signatures of a heuristic, are society's rules of thumb for successful living. Too many cooks do not always guarantee that the broth will be spoiled. And what are we to make of the conflicting advice *Look before you leap* and *He who hesitates is lost*? As with conflicting heuristics, other rules of thumb in the total context select the appropriate aphorism for use in a specific case. These pithy statements are also dated. Only recently the sayings *There is no free lunch; Everything is connected to everything;* and *If you are not part of the solution, you are part of the problem* have made their appearance. Now a decade and a half later, near age forty-five, the author of *Never Trust Anyone over Thirty* is publicly having second thoughts about his contribution. Aphorisms are social heuristics that encapsulate human experience to aid in the uncertain business of life.

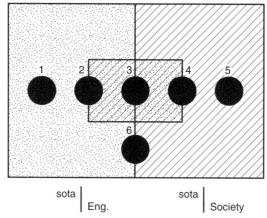

sota | Eng.   sota | Society

Figure 18 Intersection of Engineer and Society

Society solves problems, society uses heuristics, society has a sota. Some of the heuristics used by the engineer and nonengineer are the same, but each reserves some for exclusive use. Few engineers use the Ouija board, astrology, or the *I Ching* in their work, but some members of society evidently do. On the other hand, no layperson uses the Colburn relation to calculate heat transfer coefficients, but some engineers most certainly do. Therefore, the sota|$_{soc.}$ and the sota|$_{eng.}$ are not the same, but will have an intersection, as shown in Figure 18.

Here the stippled sota|$_{eng.}$ from Figure 17 has been combined with a crosshatched one representing society. For convenience, the shape representing the heuristics used by the nonengineer is taken as similar in shape to that of the engineer, but crosshatched and flipped over. A heuristic earns admission to the small rectangle of intersection of the two by being a heuristic known to both the engineer and the nonengineer. Figure 18 also includes six solid circles labeled 1–6 to indicate subsets of heuristics needed to solve specific kinds of problems. Problem 1, lying outside of the sota of society, requires only engineering heuristics; problem 2 requires some heuristics unique to the engineer combined with some from the overlap, and so on. As before, both sota|$_{eng.}$ and sota|$_{soc.}$ are composites of overlapping sotas of individuals, and therefore, the problems represented by the subsets 1–6 will often require a team effort, possibly including both engineer and nonengineer. Each of these problem areas will now be considered individually.

Problem 1 requires only engineering heuristics for its solution. Since the engineer has traditionally responded to the needs of society, few engineering problems lie in this area. The situation would be different if sota|$_{eng.}$ were replaced by sota|$_{sci.}$. Pure scientists are currently trying to decide whether time runs backward in certain particle reactions and to reconcile the theories of relativity and quantum mechanics. In each of these cases, the scientist interacts only slightly with society, and her work would be a good example of problem 1. Problems of concern only to the engineer, however, are rare.

Problems from area 2 are endemic on the engineer's drafting board. They require information from society to define the problem for solution and heuristics exclusively known to the engineer to solve them. The analysis of the engineer's notion of best given at the end of Part I serves to indicate the importance of this region and the interplay of engineer and society. The layperson can never completely understand the trade-offs necessary for the design of an automobile. She must delegate responsibility to the engineer to act in her stead and then trust the engineer's judgment. The alternative approach of restricting the engineer to problems that require no specialized knowledge—that is, those requiring no complicated computer models, no advanced mathematics, and no difficult empirical correlations—would soon grind engineering machinery to a halt. Problems such as number 2 require a joint effort in defining goals and solution strategies, but they also require heuristics unique to the trained engineer for their solution.

Some argue that we are witnessing a shrinking of this area as society disciplines the engineering profession because of disagreement over past problem solutions. No longer is it sufficient for an engineer to assert that a mass transportation system or a nuclear reactor is needed and safe. For a problem such as number 2, confidence to accept the engineer's judgment outside of the area of overlap is based, in part, on an evaluation of the engineer's performance within the area of overlap, which depends, in turn, on society's understanding of the engineering method. A simple test is in order. Ask the next nonengineer you meet: What does the word *best* mean to an engineer? How is it related to optimization theory? What is the state of the art? And what is technical feasibility? A person unable to respond satisfactorily to even these simple questions is technologically illiterate and in no position either to delegate important aspects of her life to the engineer or, more importantly, to discipline the engineer if she does not agree with the engineer's proposed solution. Given the large number of problems in region 2 and their importance, how can society afford humanists who do not have even a superficial knowledge of the major ideas that permeate engineering? What is most urgently needed is research to determine the minimum overlap necessary for a nonengineer to be technologically literate.

Problem 3 will not detain us for long. In this area of complete overlap between society and the engineer, the only dispute is over which heuristics are best to define a problem and to solve it: those common to both the engineer and nonengineer, those used by the engineer alone, or those used by one of the various subgroups of nonengineers. The politician, economist, behavioral psychologist, artist, theologian, and the engineer often emphasize different aspects of a problem and suggest different approaches to its solution. Nothing could be more different than the heuristics prayer, positive reinforcement, and Freudian psychology when it comes to rearing a child or the heuristics used by the politician, economist, and engineer when it comes to reducing the hunger in the world. Many options are available to solve the problems in area 3. Society's sota must contain effective heuristics to arbitrate between these different strategies for solving a problem. *Each of us speaks of a problem with the accent of her sota.*

Problem 4 is distinguished from previous ones in that some of the heuristics needed for an acceptable solution are not found within the sota usually attributed to the engineer in his role as engineer. Remember the San Francisco Embarcadero. Too expensive to tear down, for years it stood as a monument to a purely engineering solution that failed because the sota used by the engineers did not contain all of the heuristics that were important to society. Numerous studies show that, compared with the population as a whole, the American engineer is less well read, a better family member, more conservative politically, more oriented to the use of numbers as opposed to general philosophical positions to make a decision, and more goal oriented. These characteristics may change in the future, of course, and are surely different in different cultures, but the conclusion is inescapable. An engineer is not an average person. Therefore, when he chooses the important aspects of a problem and the relative importance of them, at times his model will not adequately represent society. The engineer may feel it is obvious that there is an energy shortage and that we need nuclear power; some members of society do not agree. The engineer may feel that it is obvious that the scientific view of the world is true; some theologians do not agree. Some engineers may feel that no book should be printed that does not contain at least one mathematical formula; those who consider the Bible, Talmud, or Koran as the ultimate guide for the salvation of their souls certainly do not agree. Only two ways of solving problems in area 4 are possible. Either the engineer can delegate responsibility for certain aspects of a design to the layperson and accept her input no matter how unreasonable it seems, or he can increase his general sensitivity to the hopes and dreams of the human species—that is, increase the overlap of his sota with that of society. But sensitizing you and me to the human condition is the responsibility of the novelist, psychologist, artist, sociologist, and historian—in short, the humanist and the social scientist. Another test is in order. Ask the next engineer you meet: What is the central thesis of behaviorism? What is the philosophical status of ethics, values, and perception? What is Impressionism's influence on art? And what is the difference between Greek and Shakespearean tragedy? An engineer unable to respond satisfactorily to even these simple questions is illiterate in the liberal arts and is in no position either to delegate important aspects of his life to the humanist or, more importantly, to discipline the humanist if he does not agree with the humanist's proposed solution. Given the large number of problems in region 4 and their importance, how can society afford engineers who do not have even a superficial knowledge of the major ideas that permeate the liberal arts? What is most urgently needed is research to determine the minimum overlap necessary for an engineer to be liberally educated.

Problem 5 requires heuristics completely outside of the expertise of the engineer and is primarily of interest to us here in determining how general the engineering method is. Since generalization of the engineering method to the universal method is the objective of a later discussion, for now, simply recall the discussion near the end of Part I. There we examined several problems posed by American presidents that would have profited from the help of an engineer. On this basis alone we can conclude that the applicability of the engineering sota is larger than usually expected. Consequently the area represented by problem 5 is much smaller than commonly supposed.

Finally, problem 6 is included to give equal time to an aberrant view of engineering. Some people, including some engineers, believe that no overlap should exist between the sotas of society and the engineer. In this view the duty of society is to pose the problems it wants solved, and the duty of the engineer is to solve them using the best techniques available. In class I call this the transom theory of engineering. Some people feel that it is society's responsibility to pitch a problem through the transom over the door to the engineer and for the engineer to pitch his solution back to society. This view fails because problems evidently exist that cannot even be defined by society without knowing what is technically feasible and because solutions evidently exist that cannot be found without knowing the value system of society. Therefore, I do not believe that many, if any, examples of problem 6 exist, or that, if they existed, they would be solvable. Despite its obvious flaws, this limited, technical view of engineering is not as rare as it should be.

Figure 18 underscores the effectiveness of the engineer's concept of a sota in the analysis of the relationship between the engineer and society. It must complete the examples intended to show the value of the concept of a sota as a tool for bringing understanding to important aspects of the engineering world. It is not giving away a secret to reveal here that I am convinced that the philosopher, like the engineer, has a sota through which she views her world. With the heuristic, the engineering method, and the sota defined, the discussion returns to our major goal for this chapter, the rule for implementing the engineering method.

PRINCIPAL RULE OF THE ENGINEERING METHOD

Defining a method does not tell how it is to be used. Method refers to the series of steps that are taken along the way to a given end. We now seek a rule to implement the engineering method. That is, we seek the steps "along the way" to an engineering solution to a problem. Since every specific implementation of the engineering method is completely defined by the heuristic it uses, this quest reduces to finding a decision rule or heuristic that will tell the individual engineer what to do and when to do it. Remembering that everything in engineering is heuristic; no matter how *clearly* and *distinctly* it may appear otherwise, I have found I have a sufficient number of rules to implement the engineering method with only one, provided that I make a firm and unalterable resolution not to violate it even in a single instance.

> The *rule of engineering* is in every instance to choose the best heuristic for use from what my personal sota takes to be the sota representing best engineering practice at the time I am required to choose.

Careful consideration of this rule shows that the engineer evaluates his actions against his personal perception of what constitutes engineering's best world instead of against an absolute or an eternal or a necessary reality. The engineer does what he feels is the most appropriate thing to do measured against this norm. In summary, at any given instance the engineer uses the heuristic that represents his "best bet" as to what to do next, all things considered. But what else could he (or any of us) do? *To know the best heuristic is to use the best heuristic.*

In addition to implementing the engineering method, the Rule of Engineering determines the minimum subset of heuristics that are needed to define the engineer. Recall that in Figure 16 the sotas of three engineers, A, B, and N, overlapped in a small rectangular subset that included the heuristics they shared. If, instead of three engineers, all engineers in all cultures and all ages are considered, the overlap would contain those heuristics absolutely essential to define a person as an engineer.

This intersection will contain only one heuristic, and this heuristic is the rule just given for implementing the engineering method. While the overlap of all modern engineers' sotas would probably include mathematics and thermodynamics, the sotas of the earliest engineers and craftsmen did not. While the sotas of primitive swordsmiths included the heuristic that a sword should be plunged through the belly of a slave to complete its fabrication, the sotas of modern manufacturers of epées do not. What heuristic must exist in the sota of a person who is said to engineer the divorce of his daughter? What heuristic must exist in the sotas of the clergy in Iran who engineered the firing of the president? In all of these cases, the Rule of Engineering, *Do what you think represents best practice at the time you must decide*, and only this rule, must be present. With this exception, neither the engineering method nor its implementation prejudices what the sota of an individual must contain for him to be called an engineer.

The goal of Part I of this discussion was to describe the situation calling for the help of an engineer. The goal in this chapter has been to describe how the engineer responds when he encounters such a situation. If you desire change, if this change is to be the best available, if the situation is complex and poorly understood, and if the solution is constrained by limited resources, then you too are in the presence of an engineering problem. What human has not been in this situation? If you cause this change by using the heuristics that you think represent those that are the best available, then you too are an engineer. What alternative is there to that? *To be human is to be an engineer.*

To say that all humans are engineers is not, however, to say that they are all equally good ones or that they are equally proficient in solving all kinds of engineering problems. No more than I would expect the civil engineer to be good at the design of a chemical plant would I expect the layperson, as engineer, to be good at airplane design. The sota available to each is different. No more than I would expect a bridge to be the same as a distillation tower would I expect a divorce to be the same as an airplane wing. Engineering design does not refer either to the specific sota used or to the artifact created. It refers to a strategy, an activity, a process for causing a change or transition—that is, to a method. Whether or not a person is designated a good engineer in a specific case depends on the previously given Rule of Judgment: *Evaluate an engineer or his engineering design against the sota that defines best practice at the time the design was made.* The prince who engineers the divorce of his daughter will fail the civil engineer's test as to what represents best engineering practice in civil engineering. The prince, although nonetheless an engineer, is hardly the one to call on to build the bridge near your home.

But to say that all humans are engineers is to say that a method exists for creating the world you would have. I ask you again: Would you establish global peace or squelch an impending family feud? Would you improve your nation or modify your local government? Would you feed the one quarter of the population that goes to bed hungry or decide what to serve for dinner? What is your fear: the mental health of our species, unbridled terrorism, genetic engineering, pollution, overpopulation? The most incisive method for causing change, any change, is now in place. Whether or not the specific change you desire occurs depends in large measure on your sota—on the effectiveness of the heuristics you know and use when you engineer your private utopia.

Some Heuristics Used by the Engineering Method

O ur attention now shifts to making our definition of the engineering method—

The *engineering method* is the use of heuristics to cause the best change in a poorly understood situation within the available resources—

more precise and succinct. As it now stands it is far too insipid for my taste, and we must now develop a second cut at the problem. To take a more aggressive posture, I will need to compel belief in the strong claim that the engineering method and the use of engineering heuristics is an absolute identity. After a brief consideration of the nature of the heuristic method and our argument to establish this identity, we will show how ubiquitous an engineer's use of heuristics really is by sampling a smorgasbord of engineering heuristics. I leave to others the gargantuan task of compiling an exhaustive list of available dishes and content myself with the selection within my more limited reach. I hope the philosopher will not find the cuisine too specialized or meager and avoid our table.

The objective of Part III is to examine competing definitions of the engineering method. This will serve as a sorbet course, to clean the palate so to speak, before continuing to the heartier fare in Part IV, where the engineering method will be generalized to a universal method. What we will find is that none of these alternate definitions is absolute, and, therefore, each is appropriately included in this part as an additional engineering heuristic.

The final objective is to reexamine the definition of the engineering method in the light of the progress we have made and to put it in a final compact form. This analysis will require us to accept some astonishing conclusions, but it will also prepare us for the more challenging analysis to come later. At this time, I hope that the engineer does not find the meal too fanciful for his taste and leave the banquet.

The caution at the beginning of the last chapter remains in effect. Although the material given here appears overly specific to engineering, it is intended to serve as a model for our coming considerations. In the present case, however, there is a second, more utilitarian, reason for reading this section carefully. At the end of the previous part we suggested that nearly all of us would wish the world were different in one aspect or another. The heuristics given here, along with the heuristics to help allocate resources that we discussed earlier, are an excellent first cut to create the world you would have if you have not developed more effective heuristics of your own.

DEFINITION OF ENGINEERING DESIGN (BIS)

In the light of what has gone before, I believe we can take the claim that engineers use heuristics from time to time as established. Our previous considerations have analyzed the type of problems the engineer solves and how he goes about solving them. By specific examples of well-documented engineering heuristics used to allocate resources, we have undeniably shown that, at the very least, engineers think they use heuristics, and it would seem wise to take them at their word. Surely it is not too hard to believe that engineers use heuristics on occasion.

Now our job is to generalize this simple claim and establish that not only do engineers use heuristics, but that that is all they do. In other words, the use of heuristics and the engineering method is an absolute identity. From the moment the engineer dons his hard hat until he removes it, his attitude toward resource allocation, risk-taking, problem solution, in fact, toward everything he does, is based on a heuristic. This is the extreme form in which we intend the definition of engineering method. Now let us look at how we will establish this fact.

THE HEURISTIC METHOD

No claim is being made that the word *heuristic* first appeared in conjunction with the engineering method or that either the heuristic or its application to solving particularly intractable problems is original with me. Some historians attribute the earliest mention of the concept to Socrates about 469 B.C., and others identify it with the mathematician Pappus around 300 A.D. Among its later adherents have been Descartes, Leibnitz, Bolzano, Mach, Hadamard, Wertheimer, James, and Koehler. In more recent times, Polya has been responsible for its continued codification. A survey of this literature shows that the study of the heuristic is very old. But as old as this research is, the use of heuristics to solve difficult problems is older still. Heuristic methods were used to guide, to discover, and to reveal a plausible direction for the construction of dams, bridges, and irrigation canals long before the birth of Socrates. Can there be any doubt that Neanderthal man used them when he engineered the final resting place for his loved one and painstakingly chipped the flints he poured into the grave?

The dominant themes of this research are adequately captured in the modern definitions of the heuristic and the heuristic method. For example, in one dictionary we read that the word comes from the Greek word *heuriskein* meaning to discover and that it applies "to arguments and methods of demonstration which are persuasive rather than logically compelling." Our use in engineering represents a significant, original departure from and strengthening of this traditional notion of the heuristic method put forth by the individuals cited earlier and by this definition.

My major problems with the traditional definition of the heuristic are twofold. First, it seems to imply that at least two different methods, a heuristic one and a logical one, exist and claims a clear distinction between them. It also appears to give us a choice as to which one to use. When we assert that the engineer uses the heuristic method, we do not admit that a different one exists to solve his problems or that he has a choice. This is fortunate, of course, because being allowed to choose between distinct alternatives at this stage would seem to doom our overall program to define universal method before it even gets started.

The additional claim that the heuristic method only persuades and the rational one compels assent is equally disturbing. If, as I assert, the engineer uses heuristics and only heuristics, it is hard to see why his designing, say, the municipal system to provide you with the life-giving fluid, water, using heuristics should be less compelling as a solution to a vital problem than the philosophers' rational arguments. Apart from these theoretical problems, we will find practical problems in making a distinction between the heuristic and the rational methods. At times, we will have difficulty trying to implement a rational approach and will ultimately have to fall back on a heuristic one.

My second major problem with the traditional definition given is that it seems to accept that a true, rational, absolute, ideal solution to a problem exists independently of the strategy used to find it. Implicit in this analysis is that the heuristic method might be able to aid in finding the solution to a particularly intractable problem, but that, in principle, this could be followed by a more scientifically acceptable strategy to find truth. To quote G. Polya in *How to Solve It*:

> Heuristic reasoning is reasoning not regarded as final and strict but as provisional and plausible only, whose purpose is to discover the solution to the present problem. . . . We shall attain complete certainty when we shall have obtained the complete solution, but before obtaining certainty we must often be satisfied with a more or less plausible guess.

The earlier example of the game of chess seems to add credence to Polya's view. Recall we admitted that a complete game tree exists for the game of chess so we can be certain that perfect play is possible. To be expedient, however, we chose the more easily implemented heuristic approach to play the game. In a similar fashion, a mathematician might accept a heuristic method that indicates the direction toward a proof, but would insist on an acceptable, rigorous mathematical demonstration to be sure. To my knowledge, no one has asked what would happen in the odd situation in which this *true* state of affairs that both the traditional heuristic and the rational methods claim to work toward turns out, itself, to be a heuristic. But this is precisely the essence of engineering. As we have already seen, the engineer's best solution to a problem is found by trade-offs in a multivariant space in which criteria and weighting coefficients are the context that determines the optimal solution. There is never an implication that a true, rational answer even exists. The answer the engineer gives is never *the* answer to a problem, but it is his engineering best answer to the problem he is given—all things considered. The Aswan High Dam is the Aswan High Dam. There is no implication that an absolute, ideal Aswan High Dam exists. Any other dam is the solution to a different problem. To sense this difference for yourself, remember the television set with one knob to control both sound and picture that we considered in Part I. Now try to imagine how you would go about setting that knob in a true, rational, absolute, ideal, Platonic way! Specifically I claim that:

1. The engineering solution to a problem has no reality apart from the heuristics used to obtain it, and
2. Everything in engineering is a heuristic.

The engineer's definition of best that we considered in Part I establishes the first claim. Now we must work to establish the second one.

NATURE OF OUR ARGUMENT

Before going to work to establish that *everything* in engineering is a heuristic, a short digression to examine the nature of my argument that this is so is in order. I would prefer to give a "logically compelling" argument that this claim is true because of the greater perceived legitimacy of logical argument in the Western tradition. But the use of the universal quantifier *everything* in my assertion presents a technical problem. Surely examining all engineering heuristics is out of the question. We simply do not have the resources to do that. If even one heuristic is left out of our consideration, there will always be a nagging doubt that perhaps, just perhaps, there is something the engineer does when he solves a problem that is not based on a heuristic. Dogma then becomes exaggeration.

The traditional way out of this problem is to use one of the philosopher's favorite strategies, called induction.* Induction refers to all cases of nondemonstrative argument, in which the truth of the premises, while not entailing the truth of the conclusion, purports to be good reason for a belief in it. Classically, induction is the process of generalization from particular instances. It is based on the feeling that in some sense Nature is uniform, that the past is a reliable guide to the future. It is that strange procedure by which scientists observe that some ostriches have long necks and then conclude that all unobserved ostriches also have long necks. The problem of induction is to justify this leap from the observed particulars to general truth. Since I cannot examine all engineering heuristics to prove my claim, I appear confined to make use of induction. But what is the fundamental nature of this strategy itself?

Being an engineer, before using induction, or any other strategy for that matter, to solve a problem, I would like to know its validity and range of applicability. I need to know this so I can apply a factor of safety to compensate for any ignorance or doubt I might have in my method of proof. What I seek before we begin, therefore, is an absolute warrant that a logically compelling argument using induction is valid. The results of many trips to the library over many years to find such assurance are somewhat disheartening.

What I find is that justification of induction is one of the classical problems in the theory of knowledge that most philosophers agree demands and at the

---

*We are not referring to mathematical induction at this time. Later we will find that mathematics and, by implication, mathematical induction will also give us cause for concern, but this is not the issue here.

same time defies a solution. We have already suggested that philosophers have doubts when it comes to the true meaning of *classification, change, definition, set,* and *ethics*. Reminiscent of these controversies is the dispute over the concept *induction*.

Induction as a Heuristic

The efforts to understand induction have been numerous, diverse, and conflicting. Some researchers have:

1. Taken the extreme step of denying that induction takes place at all,
2. Dismissed the problem by saying that induction is simply a fact of human existence,
3. Felt that an inductive argument could be converted to a safer deductive one,
4. Given up on trying to find a relationship of entailment between premises and conclusion and seek only a partial entailment,
5. Denied that inductive conclusions can even be rendered probable by the evidence in their favor, or
6. Found it totally indefensible and expunged it from any reasoning purporting to be rational.

The lack of agreement among experts is disconcerting. This is not to say, however, that technically we would be much better off if they agreed. After all, experts once agreed that the world was flat. Still, as a heuristic, agreement among experts that induction gives us true knowledge about our world would have been comforting.

The validity of an inductive argument is not an unimportant issue. Upon this rickety foundation the entire edifice of science is erected. For most people, the identification of science with induction is associated with Francis Bacon, who developed the first formal theory of inductive logic and proposed it as the logic of scientific discovery. Induction was his "new tool"* for making scientific progress. Nowadays, it is hard even to think of the enterprise *science*, if induction is disallowed. For example: We have seen the sun rise day after day in the past; how do we justify the claim that it will rise tomorrow? Or, after a thousand experiments that show that a book will fall when released or that fire will burn, what gives you the right to believe that the book you now hold in your hand will fall if released or that the fire in the fireplace across the room will burn if I put my hand into it? Of course we have no right and can only give a name to our ignorance; we call it induction.

The matter of induction is actually in even worse shape than implied here. Irreconcilable paradoxes that befuddle the layperson and expert alike plague induction. Consider two. The first, accredited to Goodman, concerns the possible existence of a rather peculiar kind of emerald, and the second, accredited to Hempel, concerns the infamous nonblack, nonraven objects.

Emeralds have many different properties. They are green, hard, and valuable. Maybe an emerald has an additional property that we will call "grue." This

---

Novum Organum.

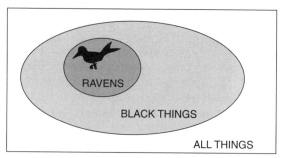

Figure 19 Hempel's Ravens

is the characteristic of being green in color up to exactly one year from today and blue thereafter. We would like to know whether emeralds are actually green or "grue" by using induction to generalize from particular instances. Suppose we examine emeralds (as many as you like) during the next year. We find all of them to be green. The problem is that all available evidence supports equally well the competing claims, "All emeralds are green" and "All emeralds are grue." As a result, what might be accepted by one expert as valid based on induction might be rejected, on equally good grounds, by another competent expert. Of course we all have the unjustified feeling that somehow our sample should take into account the time the individual objects were examined, but how is that feeling to be justified in the simple inductive argument constructed?

The second paradox asks what is to count as evidence in an inductive argument. Most people would agree that the observation of a black raven should count as evidence for the statement, "All ravens are black." This statement is logically equivalent, however, to the two statements, "All nonblack things are nonravens" (its contrapositive) and "Everything is either a nonraven or else it is black." If you happen to have the sota of a logician, this is evident by inspection of Figure 19.* The first alternate form is confirmed by the white page of the book you now hold since it is an example of a nonblack thing. The second form is confirmed by observation of the black ink used in the printing since it is an example of a black, nonraven thing. It must seem odd that by looking at the book you have in your hand and using the law of induction you have reason to believe that all ravens are indeed black. Confronted with these two examples, the expert feels as queasy about induction as do you and I. Paradoxes, inconsistencies, counterintuitive aspects, and so forth do not necessarily invalidate a concept, of course. They might just represent our current state of knowledge. Still, as long as they persist, it is probably a good heuristic to be careful about being too sure of our conclusions.

It appears that David Hume has had the last word when it comes to induction. His position is that induction cannot be logically justified. There can be no valid demonstrative argument that will allow us to infer that "those instances of which we have no experience resemble those of which we have had experi-

---

*A more formal argument that will appeal to some is given by a former student of mine, Dr. Heger: $(\forall x)(R_x \rightarrow B_x)$; R_a; $R_a \rightarrow B_a$; B_a; $B_a \vee \sim R_a$; $\sim B_a \rightarrow \sim R_a$; $(\forall x)(\sim B_x \rightarrow \sim R_x)$.

ence." The denial that Nature is uniform is not self-contradictory, and any attempt to show that induction is even probable leads to logical circularity. For instance, to prove induction would require that we produce evidence in its favor and then pass from this evidence to a general conclusion, an excellent example of induction itself.

> There is nothing in any object, considered in itself, which can afford us a reason for drawing a conclusion beyond it; and, . . . even after the observation of the frequent or constant conjunction of objects, we have no reason to draw any inference concerning any object beyond those of which we have had experience.

So says Hume; so the matter rests.

Until philosophers will give me the warrant I need that induction is absolutely justified and the results it produces are absolutely true, I am afraid I must side with the Nobel laureate Max Born when he writes in *Natural Philosophy of Cause and Chance*, "[W]hile everyday life has no definite criterion for the validity of an induction . . . science has worked out a code, or *rule of craft*, for its application. . . . [I]t is a question of faith." If a scientist of his stature has trouble accepting an inductive argument as necessarily true, it would seem a good heuristic to believe the matter is far from settled.

You can no doubt guess what I would call Born's "rule of craft." I call it the

HEURISTIC: Induction.

> Until the philosophers can solve these problems, any argument that depends on induction or Bacon's *Novum Organum* must be held to be a heuristic argument.

The engineer has never been put off, however, in solving resource-limited problems under uncertainty. What we propose is to use a heuristic method instead of a rational one to make our case compelling. Our program is therefore to compel belief in the identity between engineering method and the use of heuristics using one of the philosopher's favorite heuristics—induction.

REPRESENTATIVE ENGINEERING HEURISTICS

Some experts feel that the probability of an inductive argument being true can be increased by clever choice of examples. Likewise, if we limit ourselves to ostriches in one small locale in Africa to justify our feeling that all ostriches have long necks, we feel less certain than if we sample ostriches worldwide. If we only look at one type of engineering heuristic, drawing a conclusion about all engineering heuristics is questionable. Therefore we will divide our examples into five categories. This division is arbitrary and only for ease in showing the diversity of engineering heuristics. A definitive taxonomy of engineering heuristics must await another time and forum. Grouped together are:

1. Some simple rules of thumb and orders of magnitude
2. Some factors of safety
3. Some heuristics that determine the engineer's attitude toward his work

4. Some heuristics that engineers use to keep risk within acceptable bounds

5. Some miscellaneous heuristics that do not seem to fit elsewhere

This list should be considered as a continuation of the heuristics given in the last chapter that aid the engineer in allocating project resources. The large number of heuristics and their wide variety are important in establishing the scope of the engineering heuristic. By its extent, this list will distinguish my view from that of recent authors who limit the engineering heuristic to a routine adaptation of its traditional role in problem solution.

Rules of Thumb and Orders of Magnitude

I do not know whether I ought to touch upon the simplest heuristics used by the engineer, for they are so specific to the problem they are intended to solve that they often remain unintelligible even to engineers in closely related specialties and, hence, may not be of interest to most people. Nevertheless, to test whether my fundamental notions are accurate and to whet the appetite for what is to come, I find myself more or less compelled to speak of them. In listing these simple heuristics I do not intend to instruct in their use or even to reach an understanding of what they mean, but rather to establish their existence, their variety, their number, and their specificity.

In engineering practice, the terms *rule of thumb* and *order of magnitude* are closely related, often used interchangeably, and usually reserved for the simplest heuristics. My colleague who estimated the size of a room by knowing the order of magnitude for standard column spacing was using the kind of heuristic I have in mind, as is the civil engineer who quickly estimates the cost of a proposed highway by remembering the rule of thumb that a typical highway in America costs $1 million per mile. Both an order of magnitude and a simple rule of thumb must be considered as heuristics, of course, because neither guarantees the correct answer to a problem. Highways costing more or less than $1 million per mile certainly exist, and given the sotas of some of the avant-garde architectural engineers, I would not be surprised to find buildings somewhere with irregular column spacing. Still, both are useful to the practicing engineer whose work would be severely handicapped were these simple heuristics not to exist.

These two examples demonstrate the existence and, by implication, the importance of simple rules of thumb, but they give little indication of the variety that exists. The following group of heuristics chosen at random from the various branches of engineering will correct any chance misimpression. The

HEURISTIC: The yield strength of a material is equal to a 0.02 percent offset on the stress-strain curve

is used almost universally by mechanical engineers to estimate the point of failure of a wide variety of materials, and the

HEURISTIC: One gram of uranium gives one mega-watt day of energy

is needed by the nuclear engineer who wants to give a quick and dirty estimate of the amount of energy a power plant will generate. The chemical engineer making heat transfer calculations often assumes the

HEURISTIC: Air has an ambient temperature of 20° centigrade and a composition of 80 percent nitrogen and 20 percent oxygen,

when, in fact, the chemical plant he is designing may be located on a mountain where this rule of thumb is not exact but only an approximation. Some heuristics, for example, the

HEURISTIC: A properly designed bolt should have at least one and one-half turns in the threads

may appear rather banal, but their continued use proves their continued value. Before we had studied the heuristic in the last chapter, some of the engineer's strategies for solving problems would have seemed rather bizarre. Now we can understand what one professor at the U.S. Naval Academy means when he tells his students that if they need a quick and dirty estimate of the cost of a building, they should estimate the amount of concrete it would contain and then multiply by the cost of hamburger meat to get an approximation to the cost. What he is telling his students is that, at the present, the

HEURISTIC: The cost of building construction scales as the price of meat

is appropriate. Let me add that this same professor always continues, "If the building is to have an abnormal amount of electrical components, use the cost of sirloin instead of hamburger meat."

I doubt if these heuristics are valid decades after they were originally conceived. All of these rules of thumb should be used with care today.

This list could go on at length. As it stands, however, it is sufficient to emphasize the wide variety of engineering orders of magnitude and to demonstrate the hopelessness of ever trying to compile a complete list of the heuristics used by any one engineer, much less the engineering profession as a whole.

The engineer uses hundreds of these simple heuristics in his work, and the set he uses is a fingerprint that uniquely identifies him. The mechanical engineer knows the importance of the 0.02 percent offset on the stress-strain curve and the number of turns on a properly designed bolt, but probably has no idea how to estimate the energy release in a nuclear reactor. Likewise, the chemical engineer knows the number of plates in the average distillation tower, but does not know the strength of concrete or the average span of a suspension bridge.

In a similar fashion, the artist has a rule of thumb for the number of staples to use when she stretches her canvas and an order of magnitude for the amount of gesso to apply to it; the novelist has rules of thumb to establish her rhetorical profile and level of diction; and the chef, ones for the amount of cream of tartar needed in a chocolate soufflé. Whether lawyer, doctor, professor, or philosopher—like the engineer, the heuristics we know and use uniquely define our profession, culture, status, and era.

These simple rules of thumb represent the first category of engineering heuristics we have agreed to discuss.

Factors of Safety

One type of simple heuristic is so valuable that it is isolated here for special consideration. I am referring to the engineering heuristics called *factors of safety*.

When an engineer calculates, say, the strength of a beam, the reliability of a motor, or the capacity of a life-support system, approximations, uncertainties, and inaccuracies inevitably creep in. The calculated value is multiplied by the factor of safety to obtain the value used in actual construction.

If anyone still doubts that engineers deal in heuristics, the almost universal use of factors of safety at all steps in the design process should dissuade her from that notion. In the factor of safety we see the heuristic in its purest form: It does not guarantee an answer, it competes with other possible values, it reduces the effort needed to obtain a satisfactory answer to a problem, and it depends on time and context for its choice. In effect, it compensates for the *slop*,* that always exists in engineering design by adding an experience-determined amount of overdesign. An example will make this concept clear.

Evaluation of the appropriate size for a chemical reactor vessel requires many heuristics. At times these will include mathematical equations, handbook values, complex computer programs, and laboratory research. None of these gives an exact answer. To quote one of my former chemical engineering professors,

> Always remember that experimentally determined physical properties such as the viscosity and thermal conductivity are evaluated in the laboratory under pristine conditions. In the actual vat where the stuff is manufactured, you will be lucky if someone has not left behind an old tire or automobile jack.†

Uncertainty in the calculated value is always present, and no experienced engineer would ever believe the value he calculates would be absolutely correct. To compensate for this uncertainty, he will multiply the answer he calculates by an experience-based factor of safety. Instead of using the calculated size, he may, for example, increase it by 10, 12, or even 16 percent. In this way, many serious problems, because of the inherent uncertainty of the engineering method, are never allowed to develop.

In a similar fashion, I wonder if it would not be prudent for citizens to apply a factor of safety to the rate of acceptance of the writings of Marx or Hegel when they stake the governance of their country on these theories, for an educator to insist on a factor of safety when she applies the advice of Dewey or Spock to education, and for a person to do likewise when she develops her worldview based on the writings of Camus and Sartre.

As with all heuristics, the factor of safety depends on context. The appropriate factor to use in one situation is not necessarily the appropriate one to use in another. The engineer uses the

HEURISTIC: Use a factor of safety of 1.2 for leaf springs

when he designs the suspension system of the car you drive to the airport, the

---

*The word *slop*, although occasionally heard in engineering, does not appear in the dictionary in this sense. The engineering use of this word is undoubtedly related to the authorized definition of slop as a "loose fitting outer garment." The key idea is that an engineering calculation is a loose-fitting approximation to the appropriate design value. Related concepts are the *play* the engineer leaves between moving parts in a design and the *tolerances* he specifies in manufacture.
†Prof. Van Winkle, circa 1959.

HEURISTIC: Use a factor of safety of 1.5 for commercial airplanes

when he designs the aircraft you will take once you get there (in the military a factor of safety of 1.25 is more appropriate), and the

HEURISTIC: Use a factor of safety of 2.0 for the bolts in the elevated walkway

of the hotel where you will stay at the end of your journey. All factors of safety are not necessarily this low. In the design of a cast iron flywheel, they are often between 10 and 13. All of these numbers have come from engineering hand-books and the engineering experience of my colleagues with a time stamp of 1981. They, like all heuristics, make no pretense of being absolute, but answer only to the demands of best engineering practice at a given time. The factor of safety is our second category of engineering heuristic.

Some philosophers evidently believe that an inductive argument improves as more examples are examined. It is certainly hard not to believe that the more ostriches we find with long necks, the more probable the conclusion that all ostriches have long necks becomes. Other philosophers disagree, but at least heuristically this seems to be the case. To demonstrate the importance of the factor of safety as an engineering heuristic in engineering, you can have as many examples as you desire. The engineer has many books at his disposal, listing page after page of factors of safety, to improve our inductive argument. Pick as many as you need to convince yourself that engineers constantly use heuristics.

Attitude Determining Heuristics

Knowledge of specific rules of thumb like the ones we have just noted distinguish the engineer from the nonengineer, but this is not the only difference between the two classes of people. The modern engineer will also have more technical knowledge than the layperson will. He will know the second law of thermodynamics, be able to calculate the specific heat for steam, and know how to evaluate the Reynold's number for laminar flow in a pipe. My mother and father could not. This much is obvious. Our present interest should not be limited, however, to technical examples to find a distinction between engineer and layperson, but should also focus on those heuristics that define the attitude or behavior of an engineer when he is confronted with a problem. What does he do; how does he act? What heuristics determine the engineer's attitude toward his work and establish his unique view of the world? Our sample will include the following three heuristics as representative of the category: *Quantify or express all variables in numbers; Always give an answer;* and *Work at the margin of solvable problems.* Although some of these examples have been hinted at before, they are repeated here to demonstrate a group of heuristics that are not directed specifically at seeking the solution to a problem but are still essential and very much a part of the engineer's approach to problem solution.

The engineer is convinced that he can capture the essence of life in a net of numbers and logic. This is big game, no doubt, for so frail a snare, but no less ridiculous than the philosopher's safari armed only with words. The engineer easily falls prey to the adage of a forgotten psychometrician that "if something exists, it exists in some amount, and, if it exists in some amount, it can be measured." And so we have the engineering

HEURISTIC: Quantify or express all variables in numbers.

The use of this heuristic can at times be rather ghoulish. To quote Robert Bernero in one official report of the United States Nuclear Regulatory Commission, *Faculty Institute on Probabilistic Risk Assessment for Nuclear Power Plants,* Argonne, March, 1980:

> Some years ago the NRC took the first step by attempting the equation of the probability of loss of life with dollars. In the midst of great controversy about the effects of low level radiation on human life, the NRC was regulating nuclear reactors which released small but measurable amounts of radiation to the environment daily. The agency debated and finally chose a benchmark saying that, associating a given probability of cancer with low radiation doses, it deemed it worthwhile to further reduce releases of radioactive materials only if the cost was less than $1000 per man-rem saved. That standard was stated in Appendix I to Part 50 of Title 10 of the Code of Federal Regulations. It was and is a fascinating example of a federal regulatory body trying to use quantitative risk assessment.

This is not an isolated example. The Environmental Protection Agency has used $250,000 to $500,000 as the value of a human life in its calculations, other engineers have used $100,000 in traffic calculations, and the Federal Aviation Agency was using between $500,000 and $750,000 at the time of the TWA Flight 800 disaster.

The kind of problem attacked by the engineer and the method he uses justify this heuristic to a limited extent. We have seen that in many situations the engineer develops a model of his problem and then quantifies the axes, return functions, commensuration, and measure of goodness in order to obtain the optimum. Often this strategy works quite well, but does it always work? Do important axes exist that defy being expressed as numbers even though they are essential to good design? If they do, then we are justified in calling the engineer's use of numbers a heuristic. At present there appears to be one such neglected axis, the lumped feeling called the quality of life. Everyone wants to improve it, no one dares to try to define it, and the engineer certainly can't measure it. All too often the engineer obscures the growing tree, the flying airplane, the reddening sunset with a veil of numbers. Populations do not evolve, they grow exponentially with a characteristic time; pollution does not worsen, the amount of sulfur dioxide or number of particulates in a cubic centimeter increases; snow flakes do not swirl around corners, their kinetic energy stress tensor is anisotropic; love does not bloom, it . . . ? Never forget the San Francisco Embarcadero we studied earlier. Optimum design?

This yearning for numbers was behind the view of the engineer we have already met in the Introduction. You remember the one who felt that every book should contain at least one mathematical equation. I recently told this story to an engineering colleague expecting him to defend the engineering profession by calling this an aberrant view. But no, he answered that he thought my friend was on the right track, for even the Bible contains an implicit mathematical equation when it gives the size of Noah's ark in cubits. The engineer's craving for numbers is a powerful heuristic and determines the proper engineering attitude.

Let the engineer not forget, however, that all too often the really prized prey avoids his snare.

The willingness to decide or the willingness to give an answer to a question, any question, is another example of the proper engineering attitude. The more original and peculiar the question, the more evident the distinction between the engineer and the rest of the population becomes. The student who was willing to estimate the number of Ping-Pong balls that could be put into the classroom was obeying the engineering

HEURISTIC: Always give an answer.

This heuristic is often taught explicitly to engineering students. For example, the design of distillation towers, those familiar tall towers that dot the landscape of a chemical plant to refine petroleum products, involves the calculation of the number of plates or stages they should contain. The theoretical analysis, whose exact nature is of no concern to us now, requires a graph called a McCabe-Thiele diagram. One of my former professors once told our class in a stern voice (as he hurled a piece of chalk at any student who may have chanced to have fallen asleep):

> If you are ever in the boardroom of a large chemical company and are asked for the number of distillation plates needed to distill a material with which you are unfamiliar, guess thirteen. I'm here to tell you that, as a good rule of thumb, the average number of plates in the distillation towers in the United States is thirteen. If you know something about the McCabe–Thiele diagram for the substance in question, perhaps that it has a bump here or a bulge there, up your estimate by ten percent or lower it by the same amount. But if you admit that you have been in my class in distillation, for heaven's sake, don't say "I don't know."*

Some twenty years later (in 1982), I had occasion to ask a retired executive of a well-known chemical company this same question. Immediately he answered, "We use twenty stages as a rule of thumb; we used to use thirteen, but it is higher now."

The point of this example is clear. The engineer gives the best answer he can to any question he is asked. Of course in giving the best answer he can to a question, the engineer assumes that the person asking the question is literate in the rules of technology and understands that the answer provided is in no sense absolute but rather the best one available based upon some commonly acknowledged sota.

I confess that occasionally the engineer has exploited the public's confusion over the difference in an absolute or scientific answer and a heuristic or engineering one in dealing with problems that he would prefer to ignore. For example, if a journalist was to remind the engineering spokesmen for a large chemical plant that twenty-five of the twenty-six man-made chemicals they store in their chemical dumping ground had appeared in the ground water of the

---

*Dr. John McKetta, Jr.

neighboring area, the spokesman might well respond that twenty-five of twenty-six chemicals are not a definitive signature to identify his company as the culprit. "Theoretically," he might then continue, "if you do not know the diffusivity of the intervening soil, you cannot attribute the problem to us. And I assure you, the diffusivity of the soil in the area is not known." This engineer has conveniently misheard an engineering question as a scientific one. True enough, the exact migration of chemicals cannot be calculated without knowledge of the coefficient of diffusivity. But the technically literate reporter should sharpen her question and reel in her prey. She should now ask, "As an engineer, what is your best engineering judgment as to whether or not the twenty-five chemicals came from the only source of these man-made pollutants within miles of the town, the dumping ground of your company?" Suddenly the hooker becomes the hookee.

An additional example will demonstrate the reverse case in which an engineering answer is given, with confidence that the layperson will take it as a scientific one. When asked if nuclear reactors are safe, most nuclear engineers will immediately answer "yes." But of course nuclear reactors (or any engineering device for that matter) cannot be 100 percent safe in any absolute sense for two reasons. The first reason depends on a peculiarity of language. English has two distinct kinds of adjectives. One kind can only take values of true or false. Thus, if we ask, "Is a specific circle round?" The answer is either yes or no. There are no degrees to being absolutely round. The second kind of adjective describes a characteristic for which there are degrees or a range of values. Thus, if we ask, "Is that picture beautiful?" the answer might well be "more or less so." There are degrees of beauty. Unfortunately there are, also, degrees of safety. The second reason why a nuclear reactor cannot be considered 100 percent safe is that some possibility always exists that a factor has not been recognized in the design of a nuclear reactor that will later prove important. What the engineer means when he declares a reactor safe is that, in his view, best engineering judgment finds the probability of accident acceptably small, the risk if he has miscalculated negligible, and the need for energy so great that he recommends proceeding with the nuclear program. The reactor may be called safe from an engineering point of view, but it certainly cannot be said to be safe from linguistic or scientific ones. It is well to remember that, when asked, an engineer will always give an answer—but on his own terms.

Characterizing engineering design as the use of engineering heuristics implies that the attitude of the engineer is controlled by the additional

HEURISTIC: Work at the margin of solvable problems.

Neither problems amenable to routine analysis nor those beyond the reach of the most powerful existing engineering heuristics are included in what may properly be called engineering design. An algebra problem requiring only known, presumably noncontroversial rules of mathematics certainly would not be called an engineering design problem. Indeed, even a problem that uses only standard, well-established engineering strategies barely qualifies. They are often looked upon with derision by the professional engineer and referred to as cookbook engineering. On the other hand, a problem completely beyond the reach of even the most powerful engineering heuristics, one well outside of the sota of the engineer, would also be disqualified. Engineering design, as traditionally conceived, has no heuristics to answer the questions: What is knowledge? What is being? What is life? To qualify as design, a problem must carry the nuance of creativity, of stepping precariously from the known into the unknown, but without completely losing touch with the established state of the art. This step requires the heuristic, the rule of thumb, the best guess. If it were possible to plot all problems on a line from the most trivial to the most speculative, the engineer uses heuristics to extrapolate along this line from the clearly solvable problems into the region where the almost or partially solvable problems are found. He works at the margin of solvable problems.

Engineering is not, of course, the only profession that has heuristics governing the attitude of its practitioners. One cannot be nominated, much less confirmed, as a Supreme Court Justice without a judicial temperament; the physician is taught a bedside manner in medical school; and, I understand, the profession with only one member, the Queen of England, is taught from birth how to stop a charging bull with a regal, Victorian stare.

We have noted that the engineer is different from other people. His attitude when confronted with a problem is not the same as the average person's. The engineer is more inclined to work with numbers, give an answer when asked, and attempt to solve problems that are marginally solvable. This list completes the selection of typical heuristics that show the engineer's attitude toward problem solution. It does not, however, include all of those that could be considered or even the most important. The engineer is also generally optimistic, convinced that a problem can be solved if no one has proven otherwise, and willing to contribute to a small part of a large project as a team member and receive only anonymous glory. But the heuristics mentioned here are sufficient, I think, to indicate the presence of heuristics in the engineer's sota beyond those traditionally associated with problem solution. Any serious effort to explain the engineering method must account for these heuristics that define the engineer's attitude when confronted by a problem.

I am beginning to tire of listing engineering heuristics, and I am sure you are beginning to tire of reading them. But for induction to work its magic, a certain boredom must set in. Until we are willing to surrender and just accept that all engineering is heuristic, I feel compelled to continue.

Risk-Controlling Heuristics

Because the engineer will try to give the best answer he can, even in situations that are marginally decidable, some risk of failure is unavoidable. This does not mean, of course, that all levels of risk are acceptable. As should be expected by now, what is reasonable is determined by additional heuristics that control the size risk an engineer is willing to take. A representative group of these risk-controlling heuristics will be discussed now, including *Make small changes in the sota; Always give yourself a chance to retreat;* and *Use feedback to stabilize the design process.* This is the fourth general category of engineering heuristics we have agreed to consider.

The first

HEURISTIC: Make small changes in the sota

is important because it stabilizes the engineering method and explains the engineer's confidence in using contradictory, error-prone heuristics in solving

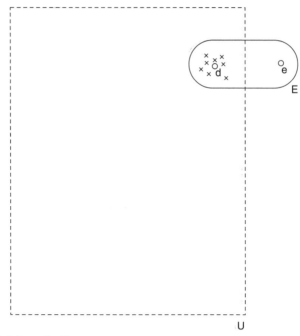

Figure 20 Validity of a Heuristic

problems, even those involving human life. Since no way exists, in advance, to be assured that a given set of heuristics will produce a satisfactory solution to a given problem, prudent practice dictates using this set only in situations that bear a family resemblance to problems for which a successful solution has been found in the past. In other words, within the hypothetical set of all possible problems, a new problem to be solved heuristically should find itself in or near the cloud of already solved problems. To illustrate this point, the sets U and E from Figure 9 are reproduced in Figure 20.

Let us assume that in the past the heuristic E has been successfully applied to problem *d* and to other problems, marked with *x*'s, that bear a marked family resemblance to *d*. In effect the engineer has built up engineering experience with E. The engineering heuristic we are now considering counsels the engineer to use E only when he can apply it to a problem located within the cloud of *x*'s in Figure 20. Under this condition the engineer is reasonably secure in using fallible, nonanalytic solution techniques. Errors do creep in. The exact position of the dotted boundary in Figure 20 is not known, and occasionally the engineer will stray across it and a design will fail. One of the most spectacular engineering failures was the Tacoma Narrows Bridge in Washington state. By oscillating with increasing amplitude over a period of days before crashing into the river below, it earned the name "Galloping Gertie." A picture of this tragedy is shown in Figure 21. When accidents happen, the engineer is quick to retreat or, as he would say, *back off* in the next use of E. By failure he has explored the range of validity of his heuristics. If you ask what is meant by "a *small* change in the sota" or "*near* the cloud of already solved problems," I will, of course, reply that the engineer has still other heuristics and extensive engineering expe-

Figure 21: Tacoma Narrows Bridge. *(Source: MSCUA, University of Washington Libraries, Farquaharson 12)*

rience with them to determine when *small* is small enough and *near* is near enough. As expected, if loss of human life is a consideration, *small* is smaller than if it is not, and engineering design becomes more conservative.

A small step does not imply no step. With the gait of a turtle, progress is made as the engineer navigates from the safety of one bank to the unknown bank on the other side of the stream using heuristics as his guide. The design of the first chemical plant to produce nylon proceeded from stepping-stone to stepping-stone, as the original theoretical idea became the bench-top experiment, the pilot plant, the demonstration plant, and finally the full-scale plant itself. This sequence, under the firm control of heuristics, allowed a safe extrapolation as knowledge gained at one step was passed to the next until the material for the blouse or the shirt you are now wearing could be produced. The engineer takes his cue from Descartes:

> As with the blind man tapping his way down an unknown path, the man makes his way carefully in the darkness. He resolves to go so slowly and circumspectly that even if he does not get ahead very rapidly he is at least safe from falling too often.

Despite its uncertainty, the heuristic method is an acceptable solution technique in part because of the stabilizing effect of this heuristic in the typical state of the art.

I remember when I first heard the next engineering

HEURISTIC: Always give yourself a chance to retreat

explicitly stated. It was in a laboratory course offered by a professor who later became a commissioner for the Atomic Energy Commission.* An expert in

---

*Professor Theos J. "Tommy" Thompson.

nuclear reactor safety, his concern was that we always had a fallback position when we built the safety systems for our nuclear reactors. Much as the computer technician stores the daily operation of his computer on a back-up tape so that operation can be resumed with a minimum of effort in case the computer fails or as the intelligent person mentally checks to see where her house keys are before she locks the door when she goes out, this heuristic recommends that the engineer allocate some of his resources to preparing for an alternate design in case his chosen one proves unworkable. Or, to use one of society's aphorisms, "Don't put all your eggs in one basket."

Competent nuclear engineers have reported that a fundamental difference in the testing philosophies of American and Russian engineers was that the former tested many design variations before settling on one, while the latter preferred to decide quickly on a reactor type, build it, and then try to make it work. Since a reactor has many components (fuel, coolant, moderator, reflector, and shield) and since many choices exist for each of these components, a large number of different reactor types are possible. To name only a few, engineers have designed pressurized water, boiling water, heavy water, homogeneous aqueous, molten plutonium, molten salt, and high-temperature gas-cooled reactors. Early in nuclear history, the American engineers built a bench-top experiment, pilot plant, and demonstration plant for as many of these different types as possible. The Russian engineers, on the other hand, selected only a few reactor types early in their nuclear program and allocated their resources to them. The difference in the two programs is shown in Figure 22, with the American system at A and the Russian one at R. At A, money is allocated for a preliminary evaluation of a large number of possible reactor types, as indicated by the lower level of the pyramid. As the evaluation proceeds, the remaining resources are funneled to the most promising concepts in an ever-narrowing manner as the engineer seeks the one best design. At R, the Russian plan calls for a much earlier choice of reactor type and allocation of all resources to it. A careful comparison reveals that both heuristics have advantages and neither can be rejected out of hand. If the design engineer can be reasonably certain that his initial choice is near optimum, or that all choices are equally desirable, the Russian system, by requiring fewer resources to reach the design objective, is clearly preferable. It does not, however, offer a chance to retreat. If the chosen reactor type proves to be physically unrealizable or economically unsound during the design process as indicated by the X at R', the Russian engineer must begin again at the beginning, as indicated by the dotted square. The American approach, at A', is more extravagant with resources, but the extra time and money invested in design alternatives can contribute in two ways. First, it assures that the final design is nearer to the optimum choice. Second, it allows a retreat to a lower level if the first choice is blocked. In a complex, unknown system, the possibility of retreat to a solidified information base will often pay dividends.

A related formulation of the rule of thumb that an engineer should allow himself a chance to retreat recommends that design decisions that carry a high penalty should be identified early, taken tentatively, and made so as to be reversible to the extent possible. This rule is so important that in some cases it is formally introduced into the design process. In the present nuclear licensing procedure, a company can get a Limited Work Authorization (LWA), which allows construction work to begin before final approval of the reactor is obtained. With this permit, only work that is reversible can be performed. An LWA is particularly important in the North, where the weather limits the time that evacuation

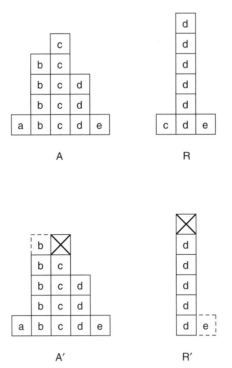

Figure 22 Russian vs. American Engineers

operations are possible. By always allowing himself a chance to retreat, the engineer has much in common with my wife, who wisely always holds an alternate in reserve when she intends to serve a temperamental dessert such as the chocolate soufflé.

The engineering term *feedback*, which is at the heart of the

HEURISTIC: Use feedback to stabilize engineering design,

must be understood before its ability to control the risk associated with the engineering method can be fully appreciated. Although crude engineering devices with feedback characteristics date from the water clock of Ktesibios in the third century B.C., the theoretical analysis of feedback is barely sixty years old. It is, therefore, surprising to find that the nonengineer is vaguely familiar with this concept and uses the term frequently, although admittedly often in an imprecise way. The teacher, lecturer, and artist seek feedback from their audiences as to whether their performances are acceptable. The biologist tells us that feedback maintains the body temperature at 98.6° Fahrenheit. The physiologist tells us that feedback maintains the body erect. The economist tells us that feedback causes the price-wage spiral. The politician tells us that feedback gone awry has caused the arms race between nations. Additional examples of the term *feedback* used by the layperson outside of an engineering context are not difficult to find.

Although the theoretical analysis of intricate feedback systems is one of the most important, sophisticated, and beautiful areas of modern engineering, because it requires complex mathematics and advanced modeling techniques, its

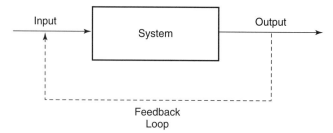

Figure 23 Feedback

precise analysis must remain beyond the scope of this discussion. Brief consideration of the effectiveness of feedback in stabilizing a complex system is essential for our work, however.

Feedback is the arrangement of any system, whether electrical, mechanical, or biological, such that the output affects the input. In Figure 23, a system, whose exact nature is immaterial, is represented by the box. On the left, it has an input; on the right, an output. Ignore for the moment the dashed line marked *feedback loop* and concentrate on the system itself. It could be a television set with the knob settings as the input and the picture and sound as outputs. It could be an automobile with the position of the steering wheel and the accelerator as inputs and the position on the road and speed as outputs. Or a system could be a room, including its furnace, for which the input is the control that determines how much heat the furnace will produce, and for which the output is the heat generated or the temperature of the room. We will consider a system as a device for transforming a change in one quantity (the input) into another one (the output). Now, to create a feedback system, we must introduce the dashed line in Figure 23. That is, we must allow the output to affect the input.

Our first example of a system with feedback, that of the thermostat that maintains the temperature of a room at the desired level, is so frequently used as an example of a feedback system that it qualifies as a genuine engineering cliché. The thermostat is a device capable of measuring the temperature of a room, comparing it to a temperature that you have previously selected (theoretically called the set point), and changing the furnace control. If the room temperature is too low, the thermostat will demand more heat by increasing the heat control; if the room is too hot, it will turn the furnace setting to a lower value. In effect, the output (room temperature) affects the input (furnace setting), and we have a stable feedback system. Identification of a feedback loop is sometimes a bit tricky. When you drive an automobile, you are yourself the feedback loop keeping the speed at the desired level. By sensing how fast the car is going, you can change the input setting of the accelerator until the speed is acceptable. If you doubt the importance of a feedback loop in keeping a system stable, try doing without one by driving blindfolded. As additional examples, you maintain your body erect and you can pick up a pencil by evaluating the deviation of your body position from its desired location and correcting the input by signals to the muscles. All of these are examples of simple systems that depend on feedback for their stability.

The engineer's concept of a simple feedback system in place, we can return to the heuristic with which this section began: the suggestion that the engineer use feedback to stabilize engineering design. To see the parallel between a feedback system and engineering design, replace the word *system* in Figure 23 with the words *engineering method*. The input will now be the sota of the engineer, and the output the results of his efforts. Earlier, we found that the system to transform this input into this output consisted of only one rule: *Choose the heuristic for use from what your sota takes to be the sota representing the best engineering practice at the time you must choose.* If this rule were all there were to engineering, engineering would hardly be a stable human activity. Failure would be rampant, and one success would not breed another. But instead, the output affects the input and a feedback loop is established. The success or failure of the engineer's effort is fed back to modify the heuristics in the engineer's sota. For me, the existence of this feedback loop will forever be enshrined in the sardonic remark of a colleague to whom I had just shown the film of the catastrophic collapse of the Tacoma Narrows Bridge. As he walked away, he said with a shrug, "Well, we'll never build one like that again."

If a bridge falls, films of the failure are studied, models of the bridge are tested in wind tunnels, and competing methods of calculation are examined to see which most accurately predicted the problem. As a result, the sota of bridge design changes. Stable engineering design requires this feedback.

Engineering is a risk-taking activity. To control these risks, the engineer has many heuristics in his sota. For example, he makes only small changes in what has worked in the past, tries to arrange matters so that if he is wrong he can retreat, and feeds back past results in order to improve future performance. Any description of engineering that does not acknowledge the importance of these three heuristics and others like them in stabilizing engineering design and, in effect, making engineering possible is hopelessly inadequate as a definition of engineering method.

Miscellaneous Heuristics

Although they do not seem to fit under any of the previous categories and you are surely tiring of this review of typical engineering heuristics, we cannot go on to more serious matters without including a brief account of three additional heuristics because of their importance and sometimes controversial nature. Each of these miscellaneous heuristics: the

HEURISTIC: Always make the minimum decision,

the

HEURISTIC: Break complex problems into smaller, more manageable pieces,

and the

HEURISTIC: Design for a specific time frame,

deals with a different subject. The first concerns flexibility; the second, specialization; and the last, planned obsolescence.

Consider the first of these. A decision, any decision, limits the scope of possible action. By its nature, it eliminates options and decreases future flexibility. Once the decision was made to develop the internal combustion engine for the automobile, retreating to consider the rotary, Wankle engine became more difficult. Once the decision was made to cool fast nuclear reactors with the element sodium, retreating to consider a gas-cooled reactor became nearly impossible. Therefore, at each stage in design, the engineer should delay his important decisions while accumulating additional information. When this rule is combined with the earlier one that an engineer "always gives an answer," a basic engineering decision policy emerges: *Delay a decision as long as possible, but once it must be made, make it courageously, based on the best available information.*

Personally I feel that philosophy would profit from a consideration of this heuristic when it comes to trying to determine the correct epistemological status of its basic concepts. We have repeatedly expressed our concern that *classification, change, definition, set, ethics,* and more recently *induction* appear in dispute within the intellectual community. Yet we still use them very effectively every day as we converse. What harm is there in labeling them as heuristics instead of insisting that they are absolute, eternal, true, real characteristics of reality? By acknowledging the current state of the art of philosophy and considering them as plausible, helpful, reasonable, but in the final analysis unjustified, incapable of justification, and perhaps fallible, what power do we deny them? It certainly would seem that at present the minimum commitment that preserves the most flexibility is to call them heuristics—subject, if you insist, to reclassification in the future.

The second miscellaneous heuristic, the advice to divide a complex problem into smaller parts, comes from no less of an authority than the namesake of this discussion. Descartes in his book that discusses the method of rightly conducting the reason and seeking truth in the sciences would have us divide a problem into as many smaller subproblems as possible. The advice offered to the engineer, on the other hand, is to subdivide a problem only until each individual part becomes coterminous with the engineer's ability to deal with it. In other words, the engineer should divide a problem only until it becomes manageable. The difficult issue being raised is that of engineering specialization. It is worth closer examination.

An engineering project is usually so complex that an engineer working alone could not complete it. The range of heuristics that would be required is simply too broad for a single engineer to be an expert in the sota representing best practice in all needed technical areas. One of the principal reasons for subdividing a project is to allow specialists to be assigned to each of its different parts. This is shown in Figure 24 (a redrawing with different emphasis of Figures 16 and 17). In this figure the heavy circle labeled P represents all of the heuristics needed to solve a specific problem, and the lighter circles labeled A, B, and N represent the sotas of different engineers or teams of engineers assigned to its solution. It is immaterial for our purposes whether the division of P between A, B, and N is along engineering disciplines (civil, chemical, mechanical), along subject areas (heat transfer, fluid flow, thermodynamics), or along individual design elements (such as a chemical plant's distillation towers, heat exchangers, or solvent extraction units). What is important is the division of labor and the specialization implied by the lack of any one engineer knowing all of the heuristics needed.

Specialization has both advantages and disadvantages. On one hand, it increases the depth and sophistication of the heuristics available for engineering design. An expert in, say, heat transfer will keep up with the relevant literature, attend specialized conferences, and have personal contacts among the researchers in the field. As a result, he will be better able to select those heuristics

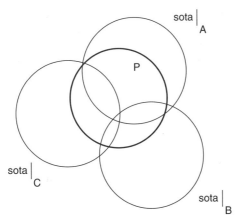

Figure 24 Engineering Specialization

most appropriate for a particular application involving heat transfer. On the other hand, subdividing a problem and assigning a specialist to each part constrain its solution. If a team of specialists in distillation has been established to design a distillation tower, retreating to consider other possible separation techniques is effectively blocked. That is, individual sotas assigned to a project limit the possible overall sota of the final project. Or in terms of Figure 24, once the circles with light borders have been selected, the maximum extent of the heavy circle is determined. A second problem with specialization is that none of the experts knows the complete axis system that determines the overall optimum solution. Each engineer is limited and captured by his own point of view. Engineers have mitigated both of these disadvantages to a large extent by using group managers with a global view of the problem and special heuristics to articulate the work of their team members. Whether or not the present *degree* of specialization is good or bad is a matter of debate, but since no two engineering sotas are ever identical, the *fact of* specialization is not in dispute.

We will discover in Part IV that the trade-off between specialization and generalization causes analogous difficulties in our search for universal method. It is immaterial whether we consider a division along intellectual discipline (engineering, philosophy, art), along work product (a bridge, the categorical imperative, *Guernica*), or along individuals (Eiffel, Kant, Picasso). A specialist's knowledge is necessary to understand the method of each, but a generalist's knowledge of all of them is necessary to achieve a synthesis.

This point will become so important later that a small, specific example will be given here using the disciplines of science, mathematics, and philosophy. At the present time, I do not expect to reach an understanding of what each example means, but only to indicate how the problem in engineering is exactly parallel to the problem in understanding the universal method. We will reexamine the specific concepts mentioned here in more detail later.

A person who is at the sota of modern physics will quite possibly be aware of the perplexities raised concerning our conventional view of reality by Einstein's thought experiment called the EPR Paradox, but totally unaware of the equally vexing problems associated with the mathematics she uses for its demonstration.

A mathematician who is at the sota of mathematics will be aware of the uncertainties about the whole realm of mathematics associated with Gödel's Proof, but totally unaware of the devastating problems raised by the skeptics in the general theory of knowledge.

Most philosophers are little better off, since they have little time after struggling with the doubts raised by Aenesidemus' Tropes concerning the theory of knowledge to worry about modern physics.

Without knowledge of the perplexities, uncertainties, and doubts raised in just these few disciplines taken together, how can we draw a conclusion that intends to encompass all of them? As we learned from Figure 24, choosing a limited set of domains, as we have just done, places additional constraints on the scope of the generalization that can be made. What about the impact on our search for universal method if we omit Hume's attack on causality, Schrödinger's wave equation, Tarski's analysis of the notion truth, and Heisenberg's Uncertainty Principle? We must solve this problem by creating new renaissance scholars and developing special heuristics to articulate the work of their less general colleagues.

Design for a specific time frame, design for a finite lifetime, or, as it is known to the nonengineer, "planned obsolescence" is a vast, complicated, and often controversial subject. It is the third, and last, miscellaneous engineering heuristic we will consider. In specific terms, it acknowledges that your new American automobile was designed to give you less than ten years of trouble-free service (the manufacturer of the Volvo claims 19.6 years for his) and then is expected to begin developing problems.* Proponents and opponents of finite lifetime design advance a multitude of often heated arguments that involve a tangle of ethical, emotional, practical, and engineering considerations on both sides of the question. We cannot stop now to referee the bout, but this much seems beyond dispute:

1. Increased product lifetime implies increased costs. In general, a toaster designed to give fifty years of service would incur more expense in its design and construction than one intended to last a shorter time. This cost is borne by the person who likes toast for breakfast.

2. Some tastes are undeniably manufactured, and as a result, the expendability of fashion replaces the durability of quality as a design criterion. This is not, however, always a negative factor. Most are happy that Marilyn Monroe did not wear a Boston Walker.

3. The sota representing best engineering practice changes with time. No one would want to drive an automobile on today's highways whose brakes, lights, suspension system, and steering were consistent with the sota of the Model-T Ford.

4. The function of some products has become obsolete and with it the desire for the product. Design of the buggy whip for a one hundred–year lifetime would certainly have been an example of overdesign.

5. From the engineer's professional point of view, the design of a product to last a long time is as rewarding as designing one with a limited life.

---

*Time stamp: 1982.

His choice between the two possible goals is made exclusively in an effort to give the public what it wants.

6. Any product that fails and is discarded before its owner is finished with it implies some waste of the materials and effort that went into its construction.

7. Some products simply last too long. Since the invention of paper in 105 A.D. by the Chinese Ts'ai Lum, the industry has worked to improve the quality and durability of writing and packaging materials. Before the recent development of biodegradable paper (incidentally, a product designed for a finite lifetime) and to a certain extent even now, the litter along the highways is testimony to their success. If the Declaration of Independence were composed today, Thomas Jefferson would have done well to write it on a candy wrapper.

Whether you are wholly for or against planned obsolescence, the engineer treats finite lifetime design as a heuristic. He determines his golden mean between the extremes of an infinite lifetime and an infinite cost and a zero lifetime and zero cost by a heuristically controlled trade-off.

We have now completed an important objective of Part III by sampling a few simple rules of thumb, orders of magnitude, and factors of safety as appetizers; by avoiding the temptation to spend too much time on the heuristics that determine an engineer's attitude toward his work and those that control the risks he takes; and by looking at a few miscellaneous heuristics. What we have found is that the range of engineering heuristics is much broader than usually recognized. With more time, we could easily extend this range even further. I, for one, am just too tired; I cannot look at another ostrich.

ALTERNATE DEFINITIONS OF ENGINEERING

We need to examine alternate definitions of the engineering method that have appeared in the literature and show why each falls short of an absolute definition and should, itself, be taken as an engineering heuristic. This study will not take long, for, unlike the extensive efforts to define the scientific method, until recently, the need to define the engineering method in a philosophically justifiable way has not been critically felt. Four competitors will be examined: the definition of the engineering method as (1) adherence to a specified morphology, (2) applied science, (3) trial and error, and (4) a problem-solving, goal-directed, or needs-fulfillment activity.

Engineering and Morphology

The most common and ambitious effort to define the engineering method by engineers is the attempt to associate it with a specific, universal structure, the so-called morphology of engineering design. Many authors have tried to define the engineering method by listing a fixed sequence of steps through which the design process is assumed to pass. For example, one recent effort gives the structure of design as analysis, synthesis, and evaluation. That is, the engineer, and if we seek a definition, presumably only the engineer,

1. Analyzes a problem,
2. Synthesizes a solution, and
3. Evaluates the results.

Polya's more classic morphology, directed at problem solution in general but adapted frequently to engineering, is understand, plan, carry out, and examine, by which is meant, the engineer must

1. Understand his problem,
2. Devise a plan to solve his problem,
3. Carry out his plan, and, finally,
4. Look back to check the solution obtained.

Perhaps the most extreme example of morphology is that of the author, who insists that to be called an engineer you must

1. Determine the specifications,
2. Make a feasibility study,
3. Perform a patent search,
4. Develop alternative design concepts,
5. Determine the selection criteria,
6. Select the most promising design concept,
7. Develop a mathematical or physical model,
8. Determine the relationship among the basic dimensions and materials of the product,
9. Optimize the design,
10. Evaluate the optimized design by extensive analysis on the mathematical model and tests of physical models, and, finally,
11. Communicate the design decisions to engineering administration and manufacturing personnel.

The basic assertion in each of these proposed definitions of the engineering method is that we know what an engineer is, what he does, and how he does it, if we can produce a list of steps in a fixed order that must be used to produce a product that is to be identified as the result of engineering.

Although these proposed structures are often helpful as heuristics and although the components of each structure often reveal many important heuristics used by engineers, structure is inadequate as a definition of design for four reasons. First, while many of the proposed structures vaguely resemble each other, most are the eccentric vision of their author. Between the two extremes just given, I found twenty-five variations on the theme before I stopped counting. From a practical point of view, a rule of thumb is needed to choose from this variety. Introducing a heuristic at this point reduces a question of dogma to one of style.

Second, the more candid authors admit that the engineer cannot simply work his way down a list of steps, but that he must circulate freely within the proposed plan—iterating, backtracking, and skipping stages almost at random. Soon structure degenerates into a set of heuristics badly in need of other heuristics to tell what to do when.

Third, none of the structures proposed to date recognizes the full spectrum of heuristics essential to a proper definition of the engineering method. Where are we counseled to make small changes in the sota? To allocate resources to the weak link? To use simple rules of thumb? The essence of engineering is not captured in the commands: analyze, synthesize, and evaluate.

Finally, to paraphrase what a scientist once said of efforts to define the scientific method as a sequence of steps, the fourth reason why structure is inadequate as a definition of engineering method is that it is highly unlikely that in actual practice engineers follow any structure proposed to explain their work. Do we really believe that Neanderthals, primitive craftspersons, early engineers, or even a team of modern engineers in practice today first *completely* understand their problem, then *completely* develop a plan, next *completely* carry out this plan, and finally examine *completely* the solution obtained? If we do not believe that they do so, are we sure that we have the right to disfranchise those who do not from engineering and say they are not engineers?

The engineer creates things that have never been. Many, perhaps most, truly creative engineers of the past cannot be shown to have followed any published morphology. Did Vannevar Bush follow a morphology on the way to his computer? What about Steve Wozniak on the way to his? Or Edwin Land and the Polaroid camera—the result of following a fixed set of steps, in a fixed, inflexible order? Medical doctors wisely use the heuristic, "At least do no harm" to guide their work. Insisting that engineers adhere to a morphology as *the* definition of engineering method when it is little more than the whim of one specific lexicographer risks causing more harm than good—at least in critical cases. It is better that a morphology be presented as a suggestion, an aid, a heuristic.

Therefore, from the perspective of this discussion, a morphology is a set of heuristics, a specific sota. It is useful as a heuristic for the novice engineer, but insufficient as an absolute definition of engineering method. As a result, instead of a definition of engineering method, we are left with the

HEURISTIC: Use a morphology to solve engineering problems.

Engineering and Applied Science

Some authors, primarily those with limited technical training, incorrectly assert that engineering is applied science. This is the second most common attempt to define the engineering method. Misunderstanding the art of engineering, they become mesmerized by the admittedly extensive and productive use made of science by engineers and elevate it from the status of one valuable engineering heuristic among many others to identity with engineering itself. On careful analysis, however, the engineer recognizes both science and its use as heuristics, albeit very important ones, to be applied only when appropriate.

The thesis that engineering is applied science fails because scientific knowledge has not always been available and is not always available now, and because even if available, it is not always appropriate for use.

Science, using the word in anywhere near its present connotation, is a relatively new human invention. Many historians credit the Ionian natural philosophers of the sixth century B.C. as its founders. Of course, before the beginning of systematic inquiry, humans had acquired a jumble of ideas about their world that was sufficient for gathering food, constructing shelter, and burying the dead. But these beliefs were characterized by superstition, imprecision, contradiction, lack of knowledge of their range of application, confused interrelationships, appeals to mystical forces, and dependence on custom, rather than on truth, for their certification. In a word, they were heuristics. As a body they defined a sota that was crassly utilitarian and tuned to answer the questions of the moment. This sota was sufficient, after a fashion, for building bridges, irrigation canals, dams, homes, and sepulchers. In it was to be found nascent engineering. Engineering, the use of engineering heuristics, clearly predates science, the assumption that the world is amenable to critical analysis. With science yet to be discovered, early engineering could hardly be defined as applied science.

What of the present day? Scientific knowledge is still unavailable for some, perhaps most, of the decisions made by the modern engineer. The design of a system to put a man on the moon could not have depended exclusively on applied science because no one had ever been to the moon before and, therefore, he could not possibly know precisely what science to apply. The exact temperature, pressure, gravitational field, and composition of the moon were unknown. Without science, how do you apply it? Yet a man placed his foot on the moon on July 20, 1969.

A second reason why engineering cannot be appropriately called applied science is that sometimes the engineer does not use available scientific knowledge that bears on his problem. We have already examined this fact in Part II when we considered the rational course of using the irrational heuristic for solving certain kinds of problems. Since resources define an engineering problem, an engineer must make his decisions within the amount he is allocated. Developing, retrieving, and applying scientific knowledge always incur cost. In some cases the engineer is so poor that he can only afford past experience, intuition, folklore, and educated guesses to solve his problems, while in others he is rich enough to afford science. The choice in each instance is dictated by his sota. It is simply not the case that the engineer uses science when available irrespective of the cost. Whether because science is unavailable or because it is too expensive, the thesis that engineering is defined as applied science must be rejected. Indeed, in Part IV we will find that, ironically, precisely the opposite relationship holds true between engineering and science.

Although we cannot say that engineering is applied science, engineers do apply science when appropriate. The rate of progress, the ease of progress, and the accuracy of progress in creating the modern world validate this heuristic. Codifying experience into theories, laws, doctrines, and principles makes our knowledge of the world more accessible and its use more economical. Measuring the tensile strength of materials that have been used in the past in bridge construction and then generalizing to unknown materials based upon an underlying theory is more efficient than the ancient engineer's slow accumulation of individual heuristics. The book you now hold in your hand would be less satisfactory if science were unavailable. Printing, composing, editing, marketing, and transporting it have depended heavily upon science. But you would most certainly be holding printed matter of some kind descending directly from the Chinese invention of movable type by Bi Sheng in the Yuan Dynasty even if the Ionian philosophers had never existed. We must admit that modern science, used as a heuristic, has fueled the machinery of modern engineering, but we should not assume it is the machinery itself.

Despite the problem the philosopher has in determining the correct epistemological status of science, the engineer perceives no serious difficulty, for he requires science only as the

HEURISTIC: Apply science when appropriate.

Engineering and Trial and Error

The problem solution strategy called trial and error has had a curious history in engineering. Undoubtedly it was first encountered as a technique for solving complex problems. Guessing the final answer (a trial) and then verifying that it is the correct one (not an error) is often the simplest way to proceed. This strategy is fairly common in some branches of engineering, notably chemical engineering, where the trial and error solution of embedded equations is almost a way of life. Perhaps following this lead, somewhat later the growth of scientific knowledge was identified by some with a trial and error procedure where a trial was called a scientific conjecture and an error, a scientific refutation. Either convinced by the esteem accorded these descriptions of science and wanting to bask in vicarious glory or aware that the engineer flounders in a world of uncertainty and at a loss to explain how he can do anything at all, many modern engineers have elevated random trial and error from one technique among many for solving difficult problems to the definition of engineering method itself.

Engineering does not, however, reduce to a simple trial and error procedure. In engineering, a wide variety of designs are not tried randomly then measured against an absolute answer, after which the failures are eliminated and the most successful retained. The problem with this analysis is that, if anything, the engineering prophet is too good. The ratio of engineering successes to total attempts is unexpectedly high. No matter how difficult an engineering task appears, somehow it always succumbs. The engineering goals of designing a supersonic airplane capable of flying faster than Mach 2, of landing a man on the moon and returning him safely to earth, and of building a power plant to exploit the nuclear fission reaction have all been established. Now supersonic airplanes, moon landings, and nuclear fission reactors all exist. Any explanation of the engineering method must explain this high success rate. Of course, a few engineering projects do fail, but these failures are always greeted with surprise. By and large, the engineer is too successful at everything he tries for simple, random trial and error to be the answer.

Instead, as we saw earlier, information derived from the completed project is returned or fed back to modify the structure of the engineering sota in a fundamental way. If past designs did not affect present designs directly and essentially, not only Gertie but also the majority of her progeny would gallop.

Any phrase that feigns to define engineering must be powerful enough to whisk us safely from the take-off of a small aircraft at Kitty Hawk on December 17, 1903, to a landing in the Sea of Tranquility on the moon on July 20, 1969, with relatively few crashes. The identification of the engineering method with the use of engineering heuristics, where one of the included engineering heuristics is "Use feedback to stabilize engineering design," not identification of the engineering method as "trial and error," can do so. The

HEURISTIC: Engineering is trial and error

is simply inadequate as a definition of the engineering method.

Engineering and Problem Solution

Other definitions of the engineering method exist. In conventional practice many are rather vaguely associated with solving a problem, attaining a goal, or fulfilling a need. Thus we read: "Design is a goal-directed, problem solving activity"; "Design is a creative decision-making process directed toward the fulfillment of human needs"; "Design is problem solving—science-based problem solving with social-human awareness"; "Design is providing physical systems for the fulfillment of human needs"; and so on. These definitions are convenient and, when speaking informally, I have used them myself. But either because they raise the troublesome question of what is to constitute a problem, goal, or need or because they commit the teleological fallacy, all such attempts at definition are in actual fact only heuristics.

A problem, goal, or need is a particularly human invention. While humans, nations, and cultures may speak of five-year plans to solve their problems, Nature (more accurately the complex of heuristics humans personify as Nature) does not seem to have seen the need for a 4.5-billion-year plan to solve hers. Americans may feel that they have a problem in getting enough energy for their automobiles, but Nature sees only an uneven distribution of energy masquerading in a variety of forms such as kinetic, mass, and potential energy, and appears totally disinterested in what this distribution is at any given instant. Likewise, it seems a bit forced to say that a river has a problem in flowing to the sea or that a rock has one in falling. To be sure, arguing that a dog, fish, or plant has a problem in finding food or shelter seems somewhat more reasonable. But, on closer analysis, even these examples appear suspect. On the other hand, wanting the divorce of a daughter, seeking the overthrow of a president, and needing a highway seem like perfect examples of problems that beg for an engineering solution.

More accurately, a problem is not a particularly human invention, but a particular human's invention. What passes as a problem for one person may not for another. When water is diverted from rivers to cool large power plants and returned at a higher temperature, some see a problem of thermal pollution, others a blessing of thermal enrichment in which the breeding of shrimp and fish in the area is enhanced. Even such a seemingly uncontroversial problem as stopping war has its detractors since some people will always profit from the fighting and will not work to end it. Of course, our overthrown president, divorced daughter, and property owner whose land was taken by eminent domain to make way for the highway will only see a problem if someone else's problem is solved. A problem is not a problem until someone thinks it is, and she thinks it is based upon the value heuristics in her own sota. This ambiguity in knowing what is to constitute a problem in an absolute sense is the first reason why identification of engineering as problem solution is at best a heuristic.

Weakening the problem-solving aspect of engineering to avoid personal value judgments and replacing it with the more neutral definition of engineering as a goal-directed activity is not much help. In this form, engineering commits the teleological fallacy. Teleology is the study of design in Nature. That is, it is the characteristic of Nature, or natural processes, of being directed toward a specific end or purpose. In layperson's terms, it is the notion that the future can somehow affect the present. The question we must now answer is, "In engineering, does the future really affect the present, or is the goal-directed aspect of engineering only an illusion based on a lack of understanding of the sophistication of the sota of the engineer?"

Do you want a bridge? The engineer will gladly design one for you. Do you want an automobile? That too is yours. It certainly seems that the engineer works backward from a goal, that this goal influences the strategy the engineer uses to reach it, and thus that engineering is teleological.

These repeated successes at establishing goals and then achieving them might be taken as adding credence to the idea that the future can pull the present toward it. This explanation neglects the complexity of the engineering sota that makes a teleological explanation unnecessary. Two examples will demonstrate this often forgotten complexity.

Although engineering goals are certainly desirable, they are seldom the most desirable ones that could have been established. Why was an airplane that would go Mach 10, a manned landing on Pluto, or a power plant using nuclear fusion not considered? If engineering were teleological and engineering objectives were based exclusively on their desirability, I am surprised that these more advanced goals were never established and, once established, achieved.

The heuristic needed to explain the engineer's remarkable success rate is that he carefully avoids problems he knows he cannot solve. In effect, the engineer is a good prophet because he only makes self-fulfilling prophecies. In other words, the engineer chooses a project based less on its desirability than its feasibility. The sota of the engineer not only contains heuristics to cause change, but also heuristics to show him which changes he can cause. The engineer calls this heuristic the *feasibility study*. In a feasibility study resources are allocated, not with the goal of solving a problem, but with the goal of finding out if a problem is solvable. In effect, a small amount of resources are allocated to see if a project can be achieved to avoid squandering a larger amount of resources on an impossible one. This goal of determining the feasibility of an idea is achieved whether the final answer as to the idea's feasibility is yes or no. Even in a feasibility study, the engineering prophet keeps his reputation intact by only considering goals that he knows he can attain. Any engineer who was practicing in 1981 would tell you that neither an aircraft that would go Mach 10, a manned landing on Pluto, nor a power plant based on nuclear fusion were feasible at the time. He would never have dared to establish these as engineering goals and, if by chance they were established, to have expected them to be achieved.

In actual fact, the matter is far subtler and more philosophically charged. An engineer cannot even conceive of goals, much less establish or reach them, if these goals cannot be expressed in terms of heuristics in his current sota. Before 1905, the engineer could not even suggest the creation of the atomic bomb because Einstein's heuristic, $E = mc^2$, was unknown. An engineering problem is nothing but a shorthand symbol for a set of current heuristics. This set does not contain any future, presently unknown ones. *The engineering prophet is successful because he only predicts what is immanent in the present sota of the engineering profession.*

We now have an explanation of the engineer's unexpectedly high success rate that does not commit the teleological fallacy based on a fuller understanding of the engineering sota. Since the identification of engineering as a goal-directed activity does commit this fallacy, I feel justified in demoting this definition of engineering to a heuristic.

The second example supporting my belief that engineering teleology is unwarranted once the complexity of the engineering sota is taken into account is based on the ability of engineers to solve problems that are originally recognized to be unsolvable. When the design of an American commercial supersonic airplane was being considered, economic heuristics dictated that it carry at least two hundred passengers. The only known material for its outer surface that could withstand the high temperatures generated at supersonic speeds with this payload was titanium. Design continued, although at the time techniques were unavailable for welding this metal. Surely this is an example of a goal that was set without knowing essential heuristics that would be needed for its achievement. Actually it is not. Once again, the paradox results from a lack of appreciation of the complexity of the engineering sota.

Recognizing that the sota is a function of time, the engineer does not base the feasibility of a design on the sota that currently exists but on the one he thinks, heuristically, will exist when he needs it. When engineers were considering the supersonic airplane, heuristics were available to predict that within the next ten years techniques would become available to weld titanium. (I might add that these heuristics were good ones for, one decade later, it was possible to weld titanium.) The feasibility of the airplane was based on current heuristics for projecting a sota into the future.

Once again, the engineering sota is seen to be far subtler than usually supposed. It contains heuristics for solving problems, heuristics for posing feasible problems, and, as we have just seen, heuristics for determining if problems will be feasible in the future. In none of these cases is the teleological fallacy committed.

Given the ability of an engineering sota to deal satisfactorily with the problem, goal, and needs aspects of engineering as well as the illusion of engineering teleology, I see no alternative but to replace definitions that contain these concepts with the

HEURISTIC: Engineering is a problem-solving, goal-directed, and needs-fulfillment activity

or some similar formulation.

We have examined four alternative definitions of engineering method: the use of a morphology, applied science, trial and error, and the vague combination of problem solution, goal achievement, or needs fulfillment that have appeared in the literature. Since each of these may also be considered as heuristics themselves, the engineering method defined in terms of heuristics subsumes them all as proper subsets.

NATURE AS DESIGNER

From its birth with Saint Augustine as midwife, through its zenith with Newton as its mentor, to its death with the British empiricist Hume as its executioner, the appearance of design in Nature is one of the most persistent arguments for the existence of God. As I understand the assertion, if we were to find a watch on the beach, we would immediately infer the existence of a watchmaker, and, in an analogous way, we should infer the existence of God from the observation of design in the Newtonian world. This is not the time or place to sort out the problems of theology, but curiously Nature does appear to be designed not so much in the stagnate laws of Newton as in the fluid strategies of the engineer. If we claim that engineers use heuristics in design and if Nature has the illusion of design, perhaps we can use these two notions as final support for our definition of engineering.

The processes that guide evolution in Nature provide excellent examples of the heuristics that guide engineering design. In overall outline, most geneticists believe that an individual's uniqueness is determined by the genetic information contained in her chromosomes in the nucleus of the cells. Specific locations along the chromosomes called genes contain the many genetic factors that make her what she is. This genetic information is encoded in a large molecule best known by its acronym, DNA, which replicates itself endlessly through the intermediary of a similar molecule known as RNA. RNA uses an existing DNA molecule as a template and then moves and assembles a copy of the original from chemicals called bases found in the environment. Diversity is created in at least three ways: half of a father's chromosomes, in combination with half of a mother's chromosomes, determining what sort of child a couple will have; random, chance mutations caused by radiation and other means occurring in the DNA; and through a process called genetic crossover. Most biologists are convinced that these small differences in individuals allow some couples to leave more progeny and others to leave less depending upon each's fitness in the existing environment by a process called natural selection. The better the design, the better the chance of survival. Over the years the small variations so selected accumulate to produce the complex organs we see today, such as the human eye. Even this greatly simplified summary of the process of evolution parallels many of the heuristics used by the engineer to attain a stable design process.

Our definition of engineering and many of the other heuristics in engineering design are consistent with this view of natural selection. First, like engineering, evolution clearly deals with *change* since that is, in fact, what the word means. The change that results from the process of natural selection is closer to the optimum or *best* change used in engineering than it is to the one proposed by Plato since an individual's survival depends on a trade-off of many different criteria instead of progression toward an idealistic form. The giraffe could not just have developed a long neck to feed on the tops of trees, for, at the same time, its neck must not have become so long that the giraffe could not escape its predators. Likewise, the giraffe's circulation system had to evolve in a unique way to support this long neck, but this change was traded off against the need to provide oxygen to the rest of the body. Saying a particular giraffe is a good, better, or best giraffe against some static, ideal giraffe seems a bit hard to accept. Just as the Aswan High Dam is the Aswan High Dam, the giraffe you now see is the giraffe you now see. Both are the best solutions to the current problem presented to the engineer or to Nature. *Uncertainty* exists in evolution as it does in engineering. At the very least, gray squirrels could not know the effect on their survival of the red squirrels with which they compete in any reasonable sense of the word *know*. We can also be certain that if the dinosaurs knew they were going to become extinct, they did not make very good use of that information. Finally, Nature is clearly limited to the *resources* at hand. If the RNA molecule does not find the material it needs in the environment to make a new replica of the original DNA molecule, all is lost.

Nature appears to implement many of the heuristics we have previously labeled as engineering heuristics. For example, take the heuristic *Allow yourself a chance to retreat*, which Nature seems to respect by maintaining variations in a species. In the distant past one species of moths in England was light-colored so that its individuals would blend in with the whitish bark of trees in existence at the time and not be eaten by birds. A slight variation between individuals existed, however. When smoke from burning coal darkened the trees, the darker moths were selected and began to dominate the species. Much later, when the English cleaned up the smoke and the trees returned to a lighter color, the moths also returned to their lighter color. The variability of the individuals in a species allows the species to survive in the vagaries of the context in which they find themselves. In fact, biologists maintain that the millions of years of existence of the cockroach is the result of its unusually effective ability to adapt to changing environments. These examples certainly make it appear that Nature is using the same heuristics as the engineer.

All engineering heuristics do not appear to have been used by Nature, however. Search as I may, I cannot find the equivalent of the human engineer's feasibility study in the strategies used by Nature. As we saw in a previous section, the ability to investigate whether or not a problem can be solved instead of limiting ourselves to solving a problem itself is one of the most effective strategies used by modern engineers to minimize the waste of resources. Nowhere do I find a process isomorphic with the feasibility study in the theory of genetics—unless, of course, *Nature selected the engineering mind in all humans as her new wrinkle, in part, as a way of implementing the feasibility study*. What final irony if the final judgment is that human intelligence is unfeasible?

The other definitions of engineering method that we have examined do not share this correspondence with natural selection. Nature, as engineer, does not appear to use a morphology. He does not first understand the problem, devise a plan, carry it out, and finally look back to check the solution. Insisting that Nature does a patent search or communicates the design decisions to the manufacturing personnel, as some morphologies of engineering design have insisted, is even more foolish. We certainly should not personify Nature to the point of saying that he applies science, and saying "Nature follows the laws of Nature" seems a bit of a tautology. Random trial and error of complete forms like the human eye is not observed. Instead, *small changes* in the existing genome (our sota) evolve by *successive approximations*. We have ample evidence for this process in the trays upon trays of butterflies, rodents, and, yes, humans showing the slow evolution of a species in any well-equipped museum. Nature does not create a complete novel, new organism and then see if it will work. Instead, what we see is the transmission of most of the information about the offspring intact in the DNA with only minor chance variations. The concrete organism itself preserves most of the information about the next generation. Finally, Nature does not appear to set up a problem for solution, identify a need to be filled, or choose a goal to be reached. The moths in England prove that if this were so, Nature would be far more fickle in the problem he wants to solve than we would like to think. All of the present definitions of engineering with the exception of the one proposed here fail to correspond to the processes we see in Nature as he gives the illusion of designer.

I do not claim that Nature emulates the engineer or that the engineer has consciously emulated Nature. What I do claim is that effective strategies for causing the best change in an uncertain situation within the available resources exist and that both Nature and the engineer have converged on many of the same ones over time. The similitude between design in Nature and design in engineering adds credence to the definition of engineering method.

We had agreed to take the fact that engineers use heuristics on occasion as established. In view of the wide variety and diversity of engineering heuristics reviewed in this chapter, our inability to produce a definition of engineering that is not a heuristic, and Nature's agreement that using the heuristics of the engineer is a good way to proceed, it seems compelling to believe that heuristics are all the engineer uses. What engineering activity can withstand the withering attack of induction and should not be considered a heuristic?

Our definition of engineering method is then not meant to imply that the engineer just uses heuristics from time to time to aid in his work, as might be said of the mathematician. Instead my thesis is that the engineering strategy for causing desirable change in an uncertain situation within the available resources and the use of heuristics is an absolute identity. In other words, everything the engineer does in his role as engineer is under the control of a heuristic. Engineering has no soupçon of the absolute, the deterministic, the guaranteed, the true. Instead it fairly reeks of the uncertain, the provisional, the doubtful. The engineer instinctively recognizes this and calls his ad hoc method "doing the best you can with what you've got," "finding a seat-of-the-pants solution," or just "muddling through." For my part it was the chance conjunction of the desire to learn chess and a course in artificial intelligence in 1965 that originally suggested this identity. It was not until some time later, however, that I was sufficiently convinced of its truth to defend it before my colleagues.* Since that time, this research has resulted in many publications, including a monograph for the American Society for Engineering Education; has been cited in articles and books; and is quoted in some of the most popular introductory textbooks in engineering design.

From the time of Socrates to the work of Polya, the heuristic method was considered as an alternative strategy for pointing the way to the true state of affairs. The implicit understanding was always that there was a true state of affairs to point the way to and that the heuristic method was one of many that could be used. When we use the *heuristic method* in engineering, we do not admit that another method is possible, much less more desirable, or that a solution independent of the one defined by the method exists.

PREFERRED DEFINITION OF THE ENGINEERING METHOD

Now we must fulfill the third and final promise made for this chapter by reexamining the definition of engineering method for the last time and putting it in a particularly precise and succinct form. This preferred form will make it easier to generalize engineering method to universal method and allow us to understand an engineering worldview consistent with our analysis.

---

*In a presentation entitled "The Teaching of the Methodology of Engineering to Large Groups of Non-Engineering Students," Gulf-Southwest section, American Society for Engineering Education, Ruston, Louisiana, March 26, 1971.

Throughout this discussion, I have repeatedly expressed my preference for what I will now call the

HEURISTIC: The engineering method is the use of heuristics to cause the best change in a poorly understood situation within the available resources.

I do not mean to imply, however, that this choice is any less of a heuristic than those rejected, but only that it is a better one. Indeed, my present aim is to show that, unless we are extremely careful, this heuristic implies an unacceptable status for the concept of *time*. To avoid this pitfall, I must now make an extensive digression from our current consideration of engineering heuristics to consider what the correct status of the concept of *time* should be. Then we will see that this heuristic is preferable to its competitors because it captures the timelessness of the engineering activity.

Let me be honest: I now have a serious problem. I am not sure what I should do next, for it will be difficult, if not impossible, for you to accept my next point. Up until now, I have been suggesting that engineers constantly use heuristics, and I have offered several examples of them. For you to accept that the engineer *Always gives an answer, Attacks the weak link, Designs for a specific time frame, Applies science when appropriate,* and so forth is really not too difficult. All of these examples of engineering heuristics are acceptable because they do not require a large change in your present beliefs. In other words, they only nudge your sota in the direction I desire. Later, I will want to argue the philosophically more challenging position that the domain of the heuristic is larger, indeed very much larger, than originally supposed. As a model of this extended scope, in this section I want to compel belief that the concept of *time* is best considered as a heuristic. Unlike the previous examples, changing the status of time from an absolute characteristic of Nature to a heuristic one aggressively attacks a belief you hold dearly. It is now no longer a question of nudging your sota, but of shoving it.

Time as a Heuristic

Before I try to engineer a change in your concept of time, let's look at your current sota. You sense time, you use time, you believe in time in the *strongest personal terms*. Ergo, time exists. Look at the second hand on your watch now and then once again when you have finished this passage. "Surely, you don't expect me to believe that time did not flow, one second becoming the next; surely, you don't seriously mean to suggest the astonishing conclusion that the belief in time is a characteristic of a personal sota," you exclaim. Your belief in time is so intimate and so immediate that you cannot believe that I would seriously ask you to abandon it. Yet I do.

The plane geometry studied in high school, Euclidean geometry, is a mathematical model built on a specific set of axioms. Until the nineteenth century, mathematicians, scientists, and engineers were convinced in the *strongest personal terms* that this model gave a true representation of the universe, and they used it without hesitation in their calculations. This is no longer true. Although Euclidean geometry is still used as a working approximation when appropriate (a heuristic, we would say), the geometry of space is now felt (ironically, once again in the *strongest personal terms*) to be non-Euclidean. Since the Euclidean axioms were never shown to be an absolute feature of our world, from the beginning it would have been more prudent to treat both the axioms and Euclidean geometry itself as heuristics. Accepting an argument based on an unproven assumption as a heuristic is not difficult—or is it? Let's look at a final example. The engineer is convinced in the *strongest personal terms* that he creates an external world of dams, bridges, and roads. This argument is based on the assumption that his senses give him absolute knowledge about reality. He sees and feels the bridge; the bridge exists. As a result the engineer cannot bring himself to doubt this assumption. Where, though, is the proof (or indeed what should be accepted as proof) that this external world exists? Perhaps, as with the Euclidean model, the external world is best considered a heuristic.

From now on, I want to claim that time is a heuristic, and I want to dissuade you from basing your view that this is not true on personal conviction. We all sense a problem with the concept of *time*. Whether layperson, scientist, or philosopher, at one time or another we have all become puzzled when we try to imagine time stretching backward or extending forward to infinity. Why, then, do we resist so strenuously and refuse to agree with the philosopher-scientist, Ernst Mach, when he writes, "[Absolute time] has neither a practical, nor a scientific value and no one is justified in saying that he knows aught about it. It is an idle metaphysical conception?" What options do I have for inducing a transition from your present sota that characterizes time as an absolute to one that characterizes it as a heuristic? Several come to mind. I might try to persuade you to put aside your strong feelings about the absolute nature of time:

1. By giving a comprehensive bibliography of books by authors who have studied and tried to understand time, mostly to no avail
2. By suggesting a book that synthesizes the work of these authors
3. By quoting well-known authors who deny the absolute nature of time
4. By limiting our discussion to only one in-depth study of time
5. By listing the many famous paradoxes of time that confuse layperson and expert alike.

Unfortunately none of these suggestions by itself would be conclusive in causing you to change because each is too time-consuming. In addition, each must only be considered as a heuristic aiding belief since none offers an absolute proof. Let us look at the failure of each.

The first possibility recommends that at this point I include a list of well-known authors who have considered time most closely and then suggest that you stop here and study these original sources. One recent compilation, *Great Books of the Western World* edited by Hutchins, consists of eight closely spaced pages containing 959 citations ranging from the works of Plato, Locke, James, Hegel, and Descartes to the less familiar writings of Lucretius, Sterne, Plotinus, and Apocrypha. Even this extensive listing leaves unexamined more modern positions such as the one of Reichenback and Grübaum on the theory of branch systems, Prigogine on bifucations, and the equally extensive body of Eastern thought in which time is far more generally denied. It also neglects work by anthropologists who feel that they have located civilizations (the American Hopi Indians and the Australian aborigines, for example) with different intuitions of time. After a careful study of this literature, I am quite sure that its diversity, extent, and lack of consensus would force you to follow the prudent course of not being too dogmatic about your opinion as to the correct status of time. You still might believe in time's absolute nature, but your faith would certainly be shaken and you might admit your doubt and let me call time a heuristic. Unfortunately, this approach is impractical. The study of time has gone on for so long and in such depths that no one can swallow this ocean of information without special effort and few would be willing to interrupt our discussion now and begin a lifelong study of it. Still, taken as a heuristic, the existence of this voluminous body of literature by some of the most respected scholars should suggest that there is more to time than meets the senses.

With the option of a study guide unmanageable, the alternative of including a carefully researched synthesis of the literature cited in the previous list at this point in our discussion suggests itself. Books of this kind exist, but typically they run to well over three hundred pages. This strategy might reduce an ocean to a sea or perhaps even a lake, but it would still represent a healthy gulp. Inserting such a tome at this point, although intellectually honest and most certainly effective in casting doubt on *time*, would hopelessly divert attention from our present goal of establishing the nature of engineering. Again, however, as a heuristic, the availability of such books suggests that someone, somewhere considers the correct status of *time* as not quite beyond dispute, not unworthy of study, and not a waste of time.

Giving an abridged list of famous authors who have studied time, perhaps with some direct quotations, is an attractive alternative and one used in countless survey courses in freshman philosophy and summaries of Western thought. If we can find even a single scholarly, wise, well-known witness to justify our consternation about time, his existence would be a good heuristic for examining the matter further. I can find many. We note, for example, that G. E. Moore and F. H. Bradley argued that time does not exist; that Aquinas felt that arguments for the endlessness of time were only dialectical; that George Berkeley admitted freely,

> For my part whenever I attempt to frame a simple idea of time, abstracted from the succession of ideas in my own mind, which flows uniformly and is participated by all beings, I am lost and embrangled in inextricable difficulties. I have no notion of it at all.

and finally, that Augustine asked, "What then *is* time?" and gave the now famous answer, "If no one asks me, I know; if I want to explain it to a questioner, I do not know."

Although I soon grow weary of producing these arbitrary quotations taken from their justifying context, and you soon become disinterested, look at your watch, and start to skim, citing experts of the caliber of Mach, Moore, Bradley, Aquinas, Berkeley, and Augustine certainly impresses. As a heuristic to lead us to question the absolute nature of time, this alternative is not bad. While no one can infer the ocean he does not know from six drops of seawater, if the drops are concentrated enough, he certainly becomes aware of the salt.

An in-depth look at one isolated aspect of time might prove a valuable way to cast doubt on the concept. This strategy would only be successful, however, with a limited number of people. Only an expert who specialized in the chosen aspect could even appreciate what was at issue. Let me use the theory of relativity and Kant's epistemological analysis of time as two examples. Demonstrating that time is at best a risky concept to the professional physicist who has studied the theory of relativity is easy. He is already aware that common notions of time are badly flawed, that some nuclear reactions seem to make more sense if time runs backward instead of forward, and that many of his colleagues are busily searching for alternatives to time to use as the primary basis of physics. A person with no knowledge of the mathematical properties of four-dimensional manifolds, geodesics, Lorentz transformations, and the principle of general covariance has little chance of understanding, much less believing, Albert Einstein when he comments, "Before the advent of the theory of relativity it had always been tacitly assumed in physics that the statement of time had an absolute significance, i.e. that it is independent of the state of the motion of the body of reference," and then continues, "Every reference body (co-ordinate system) has its own particular time; unless we are told the reference-body to which the statement refers, there is no meaning in the statement of the time of an event." What Einstein is saying is that if you snap your finger at this moment, it seems only natural to believe that that snap indicates a universal *now* for every traveler anywhere in the universe, but this is not true. Extensive background in mathematics and physics is needed to even understand what is at issue in Einstein's comment.

Likewise, a philosopher who is a specialist in Kant is well aware of the problems that Kant sought to solve concerning time. The layperson, with no understanding of the *a priori*, the categories, and the transcendent "things-in-themselves," becomes hopelessly lost when he tries to read the Transcendental Dialectic of the *Critique of Pure Reason*. He has little chance of understanding why Kant denies an absolute reality to time and why he considers it as an "*a priori* intuition" or category used by the mind to make experience intelligible, the form of experience rather than its matter, or, in other words, as an *a priori* form of inner sense that is both "empirically real and transcendentally ideal."

For those with the requisite background, an in-depth analysis of time from the point of view of either Einstein or Kant would be sufficient to demote absolute time, at least as it is currently understood, to a heuristic. For those who lack this background, it would not. But perhaps just the existence of these great men's work can be used indirectly. If you and I have not studied time for a professional lifetime as have Einstein and Kant, how can we justify a position contrary to theirs? Surely both Einstein and Kant owned a timepiece.

I have a final alternative to change your belief in time. It strips the theoretical underpinning away from either of the two preceding views and gives a popularization of the results. As we saw earlier in the case of induction, if a concept contains paradoxes, inconsistencies, and counterintuitive aspects, perhaps we should be careful and consider it a heuristic. Consider relativity. Do you not consider the following to be startling?

1. All future events to which we may ever have access can be seen, from a suitable moving observer's viewpoint, as simultaneous with the present. In other words, simultaneity is no longer a valid concept at least in its classical connotation and, hence, the distinction between past, present, and future is an illusion.

2. If twins are standing on the earth and one leaves on a trip into space, he would return to the earth younger than his sibling, the faster and longer the twin travels the greater the difference in age.

3. If two persons are standing at the center of a large, round, rotating platform with synchronized watches and then one of them moves to the outer rim, upon his return to the center he will find that his watch is too slow with respect to his friend who remained still.

4. If an observer could travel at the speed of light and look around, he would observe that all clocks would show a fixed time. The passage of time had stopped altogether.

5. A clock on the sun would run slower than one on the earth because of the sun's large mass.

Most folks do. One way or another, in the past, each of these paradoxes of time have been verified experimentally. Similar unexpected paradoxes are evident in the philosophical notion of time. We could, for example, examine Kant's antinomy in which two antithetical arguments are produced: one to demonstrate that the world had a beginning in time; the other, of equal cogency according to Kant, to demonstrate that it had no beginning. While you try to catch your breath, I might ask, with the philosopher J. J. C. Smart, "In what units is the rate of time's flow to be measured? Seconds per . . . ?" All of these counterintuitive ideas have appeared in popularizations of the works of scientists and philosophers to befuddle the reader.

Now, please, review with me the general outline of the problem I have in casting doubt on time as an absolute. The characteristics of this problem were:

1. Time seems beyond controversy because it is intimately and immediately known. You seem compelled to believe in its existence. Your sota accepts time as an absolute.

2. The research into the nature of time is vast. Many competent, credible authors have studied, lectured, and written extensively about it from every imaginable perspective.

3. To neutralize the commonsense notion of time, familiarity with this literature would seem necessary. This would take a lifetime of study.

4. You have probably not made such a study.

5. You have very little time and, perhaps, interest in making such a study now.

My attempt to solve this problem could consist in suggesting that you:

1. Read 959 detailed citations. I doubt that you will have the interest to do this.

2. Read a carefully researched synthesis. Typically a book of this nature will have three hundred pages. I doubt that you will have time to do this.

3. Consider an abridged list of authors and their quotations. I doubt that you would believe quotations ripped from context or accept my selection as representative.

4. Study in depth one aspect of the problem such as relativity. I doubt that most people will have the background to do this or the willingness to generalize from difficulties with time in this one case to doubt about time itself.

5. Read the popularizations of time including the most startling results. I doubt that they would convince you.

If these suggestions were followed, I'm quite sure that faith in the absolute nature of time would be shaken and that you would be willing to make the minimum decision or commitment and call time a heuristic. But unfortunately each is too time-consuming, impractical, or unconvincing taken alone. On the other hand, perhaps the very existence of this list can be taken as a heuristic for casting the slightest doubt on time as an absolute.

In the Introduction to this book I acknowledged that one of the heuristics of good writing was for the author never to embrace his audience too closely. It seems that you will feel more at ease if I do not give you a bear hug. Since this is only a heuristic, it does not always apply, although I realize that I take a great risk by violating it. Now I put aside all artifice and address you as directly, strongly, and simply as I can. Einstein made the following statement: "[T]he distinction between past, present, and future is only an illusion—even if a stubborn one." How on earth can you or I possibly be completely confident in our personal feelings about time? How dogmatic? How arrogant? How egocentric? Should we not admit the slightest tinge of doubt and at least consider the possibility that Einstein knew something we do not?

Please admit your doubt and allow me to write the

HEURISTIC: Time.

Identifying time as a heuristic does not rob it of its power. Some of the most edifying, useful, powerful, or destructive devices we have on the earth were created using engineering heuristics. All this change of status for the concept of time does is admit the obvious: Time is a plausible aid or direction to our thinking, but in the final analysis, it is, perhaps, unjustified, incapable of justification, and potentially fallible. I know you are asking if I really do believe that time is a heuristic. I do. Nature does not prevent my raving to this extent. I do not want to appear primitive and foolish if a future twist or change of science was to render our concept of time as obsolete as the buggy whip in the blacksmith's shop.

Before I relate this conclusion to the definition of the engineering method, let me generalize the problem we have had with the concept of time. In Part IV, not only will I want you to accept that time is a heuristic, but I will also want you to accept that many, many other concepts such as *arithmetic, logic,* and *truth* are heuristics as well. These additional concepts will have characteristics that are identical to the ones that we have just found in the case of time. They seem so real, so true, and so absolute. If the literature of time is vast, the literature for all of these concepts is vaster still. Since the major point in our discussion depends on your acknowledging this body of information, my problem is to find convincing heuristics to compel you to believe that each is a heuristic. In these future cases, our work on time will serve as an excellent model.

This part of our discussion has taken a rather complicated and unexpected turn. Therefore, I think it wise to retrace our steps to ensure that we are on the same path. The first objective of Part III was to give examples of engineering heuristics. The second objective was to challenge the standard definitions of the engineering method and to suggest that they are only additional engineering heuristics. When an engineer goes about his business, he is situated in time at point A in Figure 25a (a redrawing of Figure 1).

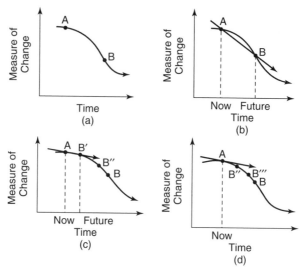

Figure 25 Time and the Derivative

As a result of his efforts at some later time the world appears as at B. If we use the heuristic that the engineering method is problem solution, the transition to point B is characterized as the solution to a problem. As we have seen, this definition is a heuristic because it depends on the present sota to define what constitutes the problem. On the other hand, if we feel that the engineering method is goal directed, we attribute to B the power to impel the engineer in its direction. When we first considered this, we identified it with its proper name *teleology*. We have seen that teleology in engineering is only an illusion caused by the complexity of the engineering sota. After the preceding discussion of time, both of these definitions are seen to have an additional disadvantage. Either the solution to today's problem or the goal that impels today's actions must lie in an assumed future. With the most celebrated scientist insisting that the distinction between past, present, and future is only a stubborn illusion, his close kin, the engineer, is left in somewhat of a quandary. We are not dealing with some idle philosophical speculation that the engineer can lightly dismiss, but with scientific theory supported by a wealth of scientific evidence. Our challenge is to formulate the definition of engineering design so that it respects time for the heuristic it is and only makes use of the present moment. After a short digression to define a mathematical object called the *derivative to a curve*, we will be in position to do this and to give a final definition of engineering method in a compact form.

Derivative to a Curve

In Figure 25b, an arrow from the point A to the point B by the shortest distance is added. A person who remembers high school geometry will immediately recognize this line as the secant. Two dashed lines are also included in this figure to emphasize the two moments in time where A and B are evaluated. One is marked *now*; the other, *future*. Let us assume that the overall change from A to B takes place in a series of smaller changes along the curve as indicated by the intermediate points B′ and B″ in Figure 25c. Between each of these new points, secants could also be drawn. We will limit our attention, however, to the first secant that shows the shortest distance from A to the first intermediate point, B′. Again, dashed lines are added to show the times when these two points are evaluated. Now I need an act of imagination that is trivial if you have the sota of a mathematician or an engineer, but next to impossible if you do not. Let us subdivide the interval A to B into more and more pieces so that each becomes smaller and smaller. As a result, B′ will get closer and closer to A until by a leap of the imagination B′ will lie on top of A; the point marked *now* will conform with the point marked *future*; and technically the secant acquires a new name. The secant is now known as the derivative, slope, or tangent to the curve at A. Although this procedure seems strange, the resultant derivative is the center-piece of a well-established branch of mathematics, the calculus. In effect, it represents an unimaginably small change in an unimaginably short period of time. For our purposes we will recognize the derivative as the direction you must go if you are at point A, and want to go to point B, and intend to follow our curve.*

The derivative to the transition path has little need for a sense of flow or passage of time. Theoretically, it is evaluated at one point in time and has no past or future. It depends on, or is evaluated at, the *now* of the world and simply indicates the direction the engineer is headed. Needless to say, different curves through the point A have different derivatives, and for each the engineer will be facing in a different direction. This local property of the derivative is certainly welcome. What I need for you to take away from this theoretical discussion is that the engineer only needs a direction in which to head at any moment in time and that that direction only depends on information in the sota of the present moment.

Reduction to Preferred Form

When we began our search for engineering method, we listed the characteristics of a situation that required the engineer to roll up his sleeves and go to work. What we found was that in each instance we were concerned with *causing change*, the appropriate notion of *best*, the problem of *uncertainty*, and the *resources* we could bring to bear on the problem. Now, we put this definition in a precise form and then in a more succinct one.

Using the derivative, I can now make the phrase "causing change," which has often appeared in the definition of engineering method, more precise. The *change* mentioned is that unimaginably small change required by the derivative. As for the correct status of the word *cause*, I will not insist upon it now. In the early moments of our discussion, we suggested that philosophers have often worried about the problem of causality. Therefore for the time being you should take the claim that "*cause* is best considered as a heuristic" on account. I will have occasion to pay my debts when we reexamine the matter later. For now, if you prefer, you may substitute *decide on, determine,* or any other similar formulation that obeys the heuristics of English grammar for the word *cause*. Of

---

*Worrying about such technical matters as the proper way to take a limit, continuity, and the treatment of the derivative as a vector at this point is clearly beyond the scope of this discussion.

necessity, the engineer must be proceeding in some direction when he causes change. The engineer's sota is the compass to help him choose the best direction to take. Therefore, we could say that the engineering method is the use of heuristics to cause the best change (to determine the most desirable derivative or to indicate the correct direction to go) in an uncertain situation within the available resource. For convenience, I will continue to use the words *cause change*, but we now know what is meant by these words.

The preferred definition of engineering method may be simplified and put in a more compact form, for its elements either describe an engineering problem situation or are themselves engineering heuristics. *Poorly understood* and the equivalent phrases used in this discussion actually refer to resources, or in this case the lack of resources. Just as with a lack of time or money, a lack of knowledge constrains a problem's solution. Therefore, this concept may be combined with the word *resource*. The engineer's best is not an absolute one, but depends upon a complex underpinning of heuristics. *Causing change* and *within the available resources* were used to describe an engineering problem situation and by implication one that requires heuristics from the sota of the engineer. As a result, the definition of engineering method as the

HEURISTIC: Use engineering heuristics

is my candidate for a final, succinct definition of the engineering method with the proviso that the phrase *engineering heuristics* is intended to include all of the specific heuristics outlined here and many others. I wonder what my definition of the philosopher's method should be? Or that of the artist? Or that of the scientist?

Justification of the Heuristic Definition of the Engineering Method

Figure 26 will help interpret the definition of engineering method and demonstrate why it is preferable to the more conventional ones that we have previously discussed. The engineer is located at point A in time, and his job is to "cause the change" to the "most desirable" final state among all of the possible final states represented by the large number of points (such as B, B', and B") in the figure. As we have repeatedly insisted, each of these final states is defined by a subset of the engineer's sota evaluated in the present. The definition of engineering method as the use of heuristics focuses attention on point A and the heuristics that define the best direction in which to go, as indicated by the arrow in the figure.

Our new definition of engineering method is a superior heuristic for at least four reasons. First, it does not require the engineer at A to know the exact final value system or even the exact final problem statement that will ultimately characterize the final point, B. The derivative, unlike the initial goal or problem, is not static as we move along the transition from A to B, but is constantly changing. The lack of information that always plagues an engineering problem suggests that during the transition from A to B, society's value system may well change and a new goal such as B' or B" may become more desirable. Second, this definition does not commit the teleological fallacy. It contains no implication that the future can somehow affect the present. Third, the proposed defi-

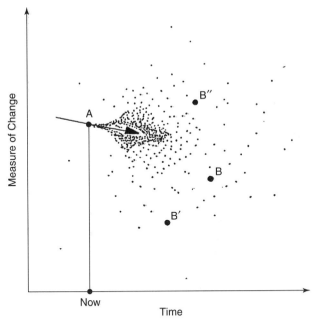

Figure 26 The Derivative and Engineering Worldview

nition is a universal definition of engineering method; that is, it is always a good heuristic in that it always describes what the engineer does. While at times an engineer is using a good heuristic when he establishes a goal or defines a problem and then sets out in its direction, often he is not. A highly hypothetical example will make this point.

Assume that a country needs more energy and has agreed on the goal of building more nuclear power stations. Since energy is needed to train people for reactor design, to mine the materials needed for its construction, and to fabricate the final station, it might turn out that to develop nuclear power, the nuclear engineer should head in the exact opposite direction and promote the construction of conventional power stations to ensure that enough energy is available for the final construction of the nuclear plant. To preserve the balance in the examples, it might be necessary for a person who is against nuclear power as an ultimate energy source to endorse nuclear power in the short term to ensure that enough energy is available to keep a society stable until his goal of an alternate energy supply could be realized. Unfortunately, establishing a goal is too often taken as a mandate to head straight in its direction. As in the two examples just given, at times this implied mandate is not a very good heuristic, and to achieve our stated goal we may appear to move away from it. But irrespective of whether establishing a goal is a good heuristic in a specific instance, whether he establishes a goal or not, the engineer is still using heuristics. In other words, "establish a goal and then try to accomplish it" may on occasion be very bad advice, but "use heuristics" is always a good heuristic to follow. The derivative or direction in which we should be headed always represents what is currently thought to be best policy. The key point is that the engineering method is primarily concerned with the transition or the strategy that safely, reliably, and predictably causes desirable change. The engineering heuristics in the present state of the art provide that direction.

The last reason why the proposed definition of engineering method is superior is that it includes the conventional ones as special cases. In a previous section, we saw that the definitions of engineering method that have appeared in the literature—*the use of a morphology, applied science, trial and error*, and *the vague combination of problem solution, goal achievement, or needs fulfillment*—are heuristics and, hence, may be included in our definition as special cases. For at least these four reasons, the engineering method is best defined as the use of engineering heuristics.

At long last we have concluded our survey of engineering heuristics including one special heuristic that defines the engineering method. This list is not intended as a complete feast of all heuristics used by the engineer or even of all the categories into which engineering heuristics may fall. My aim was to encourage you to see the heuristic behind everything the engineer does and to acquire a taste for engineering heuristics beyond the simple ones taken over from the traditional study of problem solution.

The objective of the last few sections was to dispense with time as an essential concept in engineering design. The next section does the same for the coordinate systems we have constantly used to explain engineering design.

ENGINEERING WORLDVIEW

All that remains for this part of our discussion is to propose a view of the world consistent with the assertion that everything in engineering is a function of $sota|_{eng.prof.;now}$. Our aim is to represent engineering method as self-contained without reference to an external coordinate system. This we now do by examining the widespread use of coordinate systems in engineering, science, and philosophy; by reviewing relatively recent developments in geometry; and, finally, by considering a view of engineering that is consistent with this modern analysis. Along the way we will have occasion to consider the problem of assuming that we understand one concept when in fact it depends on another unexamined concept that is laced with doubt. At this point I do not want to poach too heavily on the private preserve of either the modern physicist or the philosopher, because later I will need their help. It is always a bad heuristic to antagonize those who you will later want to enlist in your service. But I do need to refer very briefly to some of the results of their labor.

Coordinate Systems

Often engineering, modern science, and philosophy use a prior coordinate system or analogous conceptual framework to explain what they are about. At times, this framework is elevated from a crutch to aid understanding to an intrinsic component of knowledge itself.

In engineering, for example, we used a coordinate system consisting of two axes, *time* and a *measure of change*, in many of the previous figures to explain the engineering method, to derive the derivative to a curve, and to define the engineering notion of change. A point on the curve representing engineering change could be identified in terms of its location with respect to these two axes. The engineer's gibberish of secants, limits, and derivatives aimed to establish the concept of engineering design. The unavoidable feeling was that understanding the process of engineering depended on a prior unexamined commitment that such a coordinate system could be established and that the nature of the axes themselves was beyond dispute.

In modern science, one of Einstein's principal concerns was what coordinate system was the most appropriate to understand our physical world. In a way similar to the location of a point on the engineer's graph, we could talk about the location of a point in a three-dimensional space we use every day. By specifying, say, the x, y, and z coordinates in an imaginary coordinate system embedded in our conventional world, we could uniquely tell where a specific point was located. In this situation, we see all of the three dimensions as one whole, move our glance about it, and locate the point in it. When we spoke of engineering change, we seemed to feel that a special variable *time* was needed that was somehow different from the familiar ones of space. Einstein's genius was to deny a special status to time, to insist that we consider our point in a coordinate system of four dimensions, three of space and one of time, and to treat all four as equivalent. In this new situation, we could look down on the representation and see all of the four dimensions as one whole, move our glance about it, and locate information in it. Information about a point, its lifeline, or what Einstein calls its geodesic, could be grasped as a whole in terms of this four-dimensional world. The physicist's obscure gibberish concerning four-dimensional manifolds, geodesics, Lorentz transformations, and the principle of general covariance was directed at setting up the appropriate coordinate system and correctly projecting a point on the axes.

In philosophy, Kant was not concerned as much about the existence of knowledge as about the conditions under which knowledge could exist. From an engineer's point of view he made use of a very elaborate analogy to a coordinate system. Kant set up a list of schematized *categories*: quantity, quality, relation, and modality. Each of these categories was then broken down into three parts. Corresponding to quantity is unity, plurality, and totality; to quality is reality, negation, and limitation; to relation is substance-and-accident, cause-and-effect, and reciprocity; and, finally, to modality is possibility, existence, necessity. Kant held that space and time are the necessary forms of all possible experience. According to him, things-in-themselves are unknowable in themselves. They can only produce sensations in us. These sensations are ordered in our minds by means of the categories and the necessary forms, time and space. In other words, the notions of time, space, causality, and so on that we see in scientific explanation are not characteristics of the things-in-themselves but rather "the way things-in-themselves appear to us." Kant argued that these concepts simply had to be known intuitively in advance (*a priori* intuition) and that they, along with certain principles of the understanding (the categories), made experience intelligible. A bit of knowledge was possible and intelligible in terms of this framework. The philosopher's completely inscrutable gibberish concerning Kant's categories, the *a priori* intuition, and the transcendent "things-in-themselves" was directed to explaining this claim.

Whether we look at an engineer determining a point in a previously established coordinate system, Einstein insisting on the geodesic in a four-dimensional manifold, or Kant insisting that experience was only intelligible in terms of the *a priori* intuition and a complex set of categories, we find that each insists that we need a basis or coordinate system for our understanding.

Throughout intellectual history, the use of a coordinate system or an analogous framework has been a constant staple in understanding our understanding. Humans routinely try to understand one concept in terms of another, although the second is often only dimly justified or more controversial than the idea it was chosen to buttress. We have already seen that *time*, one of the axes we used in our analysis of engineering change, seems to have vanished in our hands as a firm basis of analysis upon which to build, and the other, *measure of change*, is showing signs of doubt. Often the coordinate system that is used as a basis for explanation is as unintelligible and open to dispute as the concept it seeks to illuminate.

This is not just a problem with a coordinate system, but a general problem of knowledge. Too often one concept is explained in terms of another without first establishing the validity of the latter. For example, at the very least the act of discussing this problem seems to require the assumption that language is beyond doubt and up to the task. Likewise an understanding of Kant's *a priori* intuition requires that language (and in this particular case, a translation from German), Einstein's fourth-dimensional manifold requires that mathematics, and the engineer's coordinate system requires that both language and mathematics are above suspicion and may be used with impunity in explanation. As with *time* and *change*, Part IV will insist on a tinge of doubt to *language* and *mathematics*. The subterfuge (should I say heuristic) of basing the intelligibility of one concept on the assumed intelligibility of another is far more common and far less recognized than it should be.

Although the certainty of *language* and *mathematics* must remain in dispute for the moment, at least the coordinate system that has been used to explain the engineering method is unnecessary because of new developments in geometry called *turtle graphics*. Let us turn our attention to this branch of mathematics now.

Turtle Graphics

Until now, the feeling has been unavoidable that we were standing aloof from the process of engineering, looking down on it from above, and holding it up for consideration. This was done to make engineering intelligible, but it falsifies the true situation. For the engineer solving a problem, engineering is not just an external objective of contemplation. He cannot stand apart from the process; he is embedded in it.

When most of us last studied geometry, we were captured by the legacy of Descartes. We defined our points, lines, triangles, squares, and circles in terms of a Cartesian coordinate system, and we used the analytical geometry of Descartes to describe these objects. As we have just seen, remnants of this worldview were in evidence when we used an external coordinate system to explain the engineering method. All this has changed with the creation of an alternate view of geometry called turtle graphics or turtle geometry. Our children often first meet turtle graphics when they are asked to imagine that they are sitting astride a turtle represented on a computer screen by the small triangle such as in Figure 27. They whisper instructions to their new pet with a computer keyboard using a strange artificial intelligence computer language called LOGO. All of the old geometrical objects may be redefined without the need for an external coordinate system, reference to axes, or reliance on mathematical equations

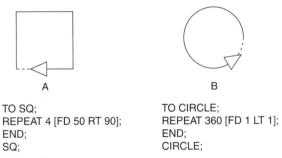

A

TO SQ;
REPEAT 4 [FD 50 RT 90];
END;
SQ;

B

TO CIRCLE;
REPEAT 360 [FD 1 LT 1];
END;
CIRCLE;

Figure 27 Turtle Graphics

using this turtle. Thrown off, at last, are the shackles of Descartes. We now look at the phenomenon itself.

An object in turtle graphics has no need for an external coordinate system but is completely defined by a specific set of instructions given to the turtle. Examples of the instructions that define a square and a circle are given to two turtles in Figure 27. The turtle at A is told to make a square with fifty units on a side by the incantation: TO SQ; REPEAT 4 [FD 50 RT 90]; END; SQ. Translated for humans, this means: If you want to make a square fifty units on a side, repeat four times—go forward fifty paces and make a right turn of 90 degrees. This procedure clearly defines a square without reference to an external coordinate system. His girlfriend at B is given the instructions that define an approximation to a circle: TO CIRCLE; REPEAT 360 [FD 1 LT 1]; END; CIRCLE. This approximation can be made more accurate by increasing the number of steps, and it approaches a circle in the limit. We no longer look down on the world, but are a part of it. All information is local; all instructions depend only on information available to the person sitting on the turtle.

Turtle geometry is not just an abstract way of doing mathematics with no counterpart in the real world. Two examples, the first of which is rather strange as befits the strange new world of our turtle, will make this point.

Turtle graphics provides a more plausible answer to how the strange, huge figures in the deserts of Peru and Chile near the Andes called the Lines of Nazca and the Giant of Atacama were made than the alternatives that have been put forth to solve the mystery. Figure 28 is one spectacular example of a huge spider that measures some 150 feet in length formed by one continuous line scratched in the earth. The menagerie comprises over three hundred figures, including a monkey, dog, condor, lizard, heron with a 900-foot-long neck, many geometrical shapes, and (at nearby Atacama) a huge giant over 390 feet long in the shape of a human dressed in crown and boots. This last figure is the largest representation of a human in the world. The total production of figures, all of which can only be seen and understood from the air, has been called one of the most baffling enigmas of archaeology. Scientists do not agree on how these large geoglyphs were made. But assuming that workers were given instructions, perhaps from a small scaled model, as they moved along from path to path is at least as plausible as other solutions currently proposed. Some insist the natives set up a huge Cartesian system. Others suggest that UFOs made the figures. Still others feel that the workers constructed them with guidance from an anticipation of the hot air balloon invented by the French many years later.

Figure 28 Figure in the Andes. *(Source: American Philosophical Society)*

Most of us have had personal experience with turtle graphics, although we may not have been aware of it. When we navigate the concourses of a large airport to find the appropriate departure gate, we rely on local signs and have no overall view of the airport. Similarly, using a map or crude instructions from a friend to drive in a strange city has no coordinate system and hides the overall contours and curvature to the space and is an example of turtle graphics. This last example is all the more startling if you are using a cellular telephone and a friend's instructions to negotiate the landmarks and streets to his house. In this case, you are the turtle receiving instructions. We will now see that these recent developments in geometry allow a particularly powerful, brief, and memorable way to represent the engineering worldview without a coordinate system.

Consistent Engineering Worldview

Turtle graphics allows a view of engineering consistent with our claim that the engineering method does not require an external coordinate system to be understandable. As shown in Figure 29, the engineer sits astride his children's cyberpet whispering instructions to it from a personal, but imperfect, copy of $\text{sota}|_{\text{eng.prof.;now}}$. Problems such as P_1 surround him on all sides. Each of these problems is defined in terms of the heuristics he knows. He tries, and seldom fails, to navigate to the ones that strike society's fancy using the best heuristics

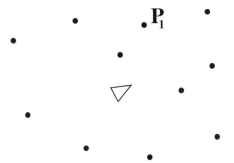

Figure 29 Engineering Worldview

he has in his sota. With this ride on the back of a very special turtle, the Cartesian model disappears.

The definition of an engineering problem is a heuristic; the definition of the engineering method is a heuristic; the rule for implementing the engineering method is a heuristic; the rule of judgment for an engineering achievement is a heuristic; and the engineer's worldview is defined by a heuristic—*All engineering is heuristic.*

Engineering, the use of engineering heuristics—with this concatenation the engineering profession is no longer obsessed with its artifacts, but becomes concerned with its art. The tranquility of mind philosophers have long sought (ataraxia) descends on the engineer as we only ask him to use the best heuristics he knows and to assure him that that is all he can be asked. The responsibility of each human, as engineer, to merit this peace becomes clear. It is for each of us to learn, discover, create, and develop the most effective and beneficial heuristics as stewards of our planet. What enormous potential is unleashed as the engineer enters his maturity and uses his method to study the engineering heuristics he will need to make a better world in the future? *The world, as we know it, vindicates the sota of past engineers; the future world, if we have it, will vindicate this new sota of the present ones.*

Until now the discussion has revolved around the engineer from the initial observation that the engineer affects our world incisively, through the discussion of the heuristic and sota, with a pause to review a selected list of engineering heuristics, and so on at last to the final definition of engineering. The discussion of feedback was a familiar landmark for the engineer and resonated with a lifetime of study. It immediately conjured up the complex issues of interlocking feedback loops, coupled differential equations, root locus plots, and the evaluation of the stability of complex systems in the minds of the modern engineer, but it was outside of the experience of most philosophers. The complex sounding discussion of the derivative to a curve we have just finished is, in reality, very common fare for the freshman engineering student studying the calculus, and mention of factors of safety and optimization theory were familiar to all engineers in all countries. Of course we have occasionally made the engineer feel ill at ease as with our suggestions that *classification, change, definition, set,* and *ethics,* and more recently *induction* and *time* were in shaky condition and by the foreboding implication that *language* and *mathematics* may not be all they seem. But so far in our analysis, it has been the engineer who has felt most at home.

The nonengineer, however, is probably in a mild state of culture shock. Now we must redress the balance and enter a new intellectual landscape where it will be the engineer who will feel homesick. My intention is to use the engineering method as a model, to extrapolate beyond it, and to create universal method. The coming discussion of Hume will be a familiar landmark for the philosopher and will resonate with a lifetime of study. It will immediately conjure up complex issues concerning causality, reality, and the theory of knowledge and will wake the rationalist from his dogmatic slumber. As a matter of fact, the complex sounding discussion of perception we will soon have is very common fare for the freshman philosophy student and the simple phrases, "I think, therefore I am" and "This is my hand" are familiar landmarks to all philosophers in all countries. In the remainder of this discussion, it will be the engineer who will experience culture shock as the heuristics we use are drawn from a different sota.

But if some philosophers have accompanied us this far, should not a few engineers make the effort to continue? The climb will be steep and the time short because the philosopher has studied the nature of his art for so much longer, and his heuristics for doing so are much more sophisticated. If you feel that the journey is not worth the view, circle the mountain where we are now located and rejoin us in Part VI where we will consider the application of method to our world. There the engineer will be a welcome antidote to the dizziness of those who have made the climb.

Finally, I must give a word of caution before we begin our journey. As I admitted in the introduction, I am an engineer. I, likewise, have felt at home with the first three parts of our discussion and will, like most engineers, feel uneasy with what is to come. As I have challenged the engineer to climb, I intend to make the effort myself, confident that the philosopher who eschews our companionship will have suffered a real loss.

Ordinarily, the engineer and the philosopher work at opposite ends of the problem spectrum. Each feels that he has an exclusive hold on the fundamental characteristics of life, and each has little interest in the other's point of view. The engineer is convinced that the philosopher, with his esoteric flights of mind, is fiddling while our civilization burns; the philosopher is convinced that the engineer, with his stubborn refusal to listen, is busy designing the best matches. Could the study of the heuristic and its application across the entire problem spectrum bridge the gap and synthesize a unified world? If the heuristic is considered in its most general form, as opposed to limiting it to the engineering heuristic, can a true gap between the problems solved and the methods used for solving them exist between the philosopher and the engineer?

IV

The *Universale Organum*

All i̲s h̲euristic. I make no small hypothesis. This modern 公* is the heuristic
公
案

from which will emerge a universal method. It is this public statement I keep
constantly before me on my desk and the one that each of us should inscribe on
the ceilings of our studies to answer the question, "What do I know?" As with
all kōans, once the claim *A̲ll i̲s h̲euristic* is truly understood, it ceases to be paradoxical and establishes a standard of judgment against which an understanding of a universal method may be tested.

This understanding begins as the engineer's world expands to encompass
each of our own worlds. The previous three parts of this discussion have examined the art that brought the engineer's vast collage of individual creations
into being. Although each of these achievements should be prized for its usefulness and beauty, the heuristics the engineer used to bring them into existence
were often crude, and the man-made world frequently appears as a collection
of random pieces stuck together in an unaesthetic jumble. Primitive though the
engineer's art is, it will serve as a model for the only art possible. To the generalization of engineering we must now turn, replacing the unrefined bits and
pieces of the engineer with the sweeping brushstrokes of the human species.
The canvas becomes the world; our school, modern.

We will investigate this generalization in six steps. First, we consider the intrinsic difficulties in even discussing this philosophical position, much less in
accepting it. Making this claim compelling is step two, where I must try to convince you that everything is, indeed, heuristic. This is an equally daunting task,
for if true, few conventional strategies are left to convince you of this fact. When
we considered the engineering heuristic, it was only a question of sampling a
few. Now it will be a question of chewing on some of the most important concepts that exist in the Western tradition. At this point the various components
of the claim *A̲ll i̲s h̲euristic* are examined to ensure that we understand their implications. We will also reduce the kōan to a more acceptable form at this time.
Next, or step three, the heuristic position is contrasted with that of the skeptic
to demonstrate how truly radical this kōan is. This step also includes an im-

---

*This cluster of Japanese characters, rendered phonetically in English as kōan,
refers to the paradoxes used by Zen Buddhists to achieve sudden intuitive
enlightenment by frustrating an ultimate dependency on reason.

pregnable defense capable of protecting those who would side with me on this matter from attack. Just as we did earlier when we discussed the engineering method, the fourth step is to reexamine the concept of a sota, to generalize and partition it, and to consider some of its most important subsets. This step also considers subsotas as structure and two important characteristics of the specific subsota each of us uses to structure his world. This consideration of a general sota is parallel to the previous one of the engineering sota, but made more precise. We will come to understand the reemergence of the Rule of Judgment and Rule of Implementation in a more general context. At last, step five establishes the long-awaited universal method. Our journey completed, Part IV closes with a brief overall review and introduces the definitive statement of the philosophical position implied by this discussion, which is given in its most intellectually satisfying form in Part V.

DIFFICULTIES IN EXPLAINING THE KŌAN

All is heuristic. I have now said it twice. The first objective of this part of our discussion is to compel belief in this paradoxical phrase. Only from a general philosophical position can a specific philosophical position that depends upon it emerge. Therefore, a careful understanding of this kōan will prove essential to a later justification of the universal method. To this end, we will examine the general difficulties that lay in the way of our understanding it and the specific difficulty of any language in trying to explain it.

Generalizing from *All engineering is heuristic* to *All is heuristic* is difficult for two reasons: The first, often thought to be serious, turns out to be minor; the second, often thought to be minor, turns out to be more serious. First, if this generalization is accepted, heuristics will have to be used to explain themselves. This problem is similar to trying to demonstrate the color white in a world in which everything is white. In fact, some philosophers argue that universal doubt and universal certainty (and I am sure they would include the notion that everything is heuristic) are untenable because of the principle of predicate contrast or sense-in-contrast. This principle holds that a predicate such as doubt and its contradictory must both apply to an understood class of statements to give sense to either. If no contrast exists between figure and ground, what is to be taken as a figure? Or in the present context, if everything is a heuristic, what distinguishes the heuristic from its contradictory? This theoretical criticism presents no problem and may be easily dismissed by remembering that if everything is a heuristic, then the principle of predicate contrast is also one.

General Difficulties

The serious problem lies elsewhere. It is the practical difficulty of using only a small subset of all heuristics to explain the basic claim that everything is a heuristic. As an illustration, Figure 30 represents the set of all heuristics. Embedded in this overall set are the subsets of heuristics that define specific areas of study such as music, the dramatic arts, science, and so forth, separated from each other by man-made boundaries. As long as we do not approach the boundary between the different fields of study too closely, examine the individual heuristics within a chosen field too carefully, or worry about the cross-pollination of one field

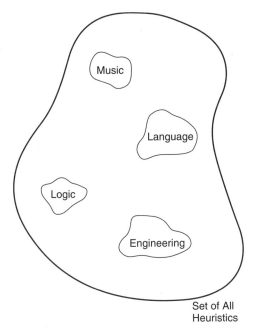

Figure 30 Set of All Heuristics

with a far distant one, each subset seems reasonably satisfactory, consistent, and complete and ultimately takes on the air of truth. Practitioners in each field are usually satisfied that their own heuristics are the measure of reality. Only when we become professionally interested in all of the individual crops of heuristics at the same time do we recognize the problems of the local farmer.

Presumably, the entire set of heuristics would be needed to give an accurate, complete description of the heuristic. That is, music, the dramatic arts, science, poetry, logic—not to mention engineering—all have a contribution to make to our understanding of the heuristic world. But for practical reasons we must hope that a fleeting feeling of the heuristic's essence can be given using a limited number of subsets within this whole. Restricted in this way, we are in the position of a colorblind person before a painting by Renoir. Perhaps we are able to discern the hand of a master, but we are forever unable to get the true impression because we cannot perceive the artist's rainbow palette. To summarize: Generalizing from *All engineering is heuristic* to *All is heuristic* is difficult, first because, theoretically, we must use heuristics to explain the heuristic, and second, and more seriously, because, practically, we are limited to only a small subset of all heuristics to fashion this explanation.

Language as Heuristic

Historically, language has assumed a privileged position in explanation. Despite the artist's insistence that art, the musician's insistence that music, the scientist's insistence that science, and the theologian's insistence that theology is capable of providing truth, when explanation is sought, most people turn to language. But since Plato counsels that "no man of intelligence will venture to express his philosophical views in language"; since the Zen Buddhist warns that "the instant you speak about a thing, you miss the mark"; and since speaking about language at all appears to beg the question by assuming that language is something that can be absolutely trusted, before we commit ourselves wholeheartedly to language's exclusive use to explain the phrase *All is heuristic* perhaps we should briefly consider language's true nature. Our first step must be, therefore, to compel belief that language is a heuristic.

If we had more time, an excellent place to begin would be with a survey of the voluminous literature on the subject. Sufficient evidence is now available to make any reasonable person cautious about accepting language's ability to explain anything with certainty. For the most part this evidence is highly technical and accessible only to the professional linguist. After studying Tarski's constructive proof that a universal language must forever remain plagued with contradiction, Whorf's view that the structure of the language one uses influences the manner in which one understands his environment, the heated dispute between Skinner and Chomsky over how language ability is acquired, and the theory that seems to show the impossibility of discussing language or the relationship between language and reality, the prudent person is forced to agree with the wit who has said, "The serious study of language appears to be a field in which words fail us." Because of the volume of information involved, changing the status of language to a heuristic using the technical literature is almost as difficult as the previous problem we would have had in changing the status of *time* and *induction* to heuristics by examining their technical literature, but not quite.

The inclination of most people who have thought about the matter at all is already to treat language as a heuristic. We have all experienced the ambiguities, miscommunications, inadequacies, and imprecision of language in our daily lives. While language is certainly helpful in explaining an idea and none of us would suggest doing without it, it is clearly fallible. Since the heuristic nature of language is not much in dispute on a practical level, instead of a literature review, this part will only offer four simple reasons to doubt for those who have never seriously considered the problems of language. We will discover that languages are different, languages vary over time, languages are unique to each person, and languages fail to adequately explain things we know to be true.

Different languages are different, fundamentally different. Therefore doubt that all languages can describe a single underlying reality would seem justified. Languages differ widely in the number of words they contain, in the relationship between the words and the concepts to which they refer, and in the basic structure and intent of words when they are used.

Languages classify items of experience differently. The number of words in different languages is strikingly different, and at times one word in one language replaces several in another. For example, Eskimo has seven words for *snow* where English has only one; Hopi has only one word *masa'ytaka* for the three English concepts *dragon fly*, *airplane*, and *aviator*; and Japanese has one symbol that means *blue* and *green* depending on context.* In fact, French has the word *sympatique* and Spanish has a similar word *simpatico* that are all but un-

---

*Compare the third Japanese character in the Japanese sentence 本は青い (The book is blue) with 山は青い (The mountain is green) Translation: SHIGETA Kazuhiro.

translatable in English and can only be understood by one who has grown up in the requisite tradition. If a language should fit the concepts it tries to explain like a glove, we have a curious assortment of zero-, one-, three-, and seven-fingered gloves from which to choose to explain the phrase *All is heuristic.*

Not only are the total number of words in various languages dramatically different and the number of words for the various concepts different, but the relationship between these words (or symbols) and the concepts they represent is also very different. For example, the relationship between the English word (a sequence of abstract letters) and its meaning is completely different from the relationship between the Chinese abbreviated picture (a group of brushstrokes) and its meaning. The word *man* relates to the notion *man* in a fundamentally different way from that in which the Chinese ideograph 人, which looks like and was derived from a picture of a man, relates to the notion *man*. An even better example is the word *like*. The English word *like* is more antiseptically related to its denotation than the Chinese ideograph 好, which combines the picture of a mother and her child.

Finally, all languages are not isomorphic. That is, they do not have the same structure. Others have noted the large number of words in English such as *because, as, although, in order to,* and *so that* that focus attention on grammar, structure, and logic. We are told that Chinese, on the other hand, has almost a total absence of grammar and organizes uninflected ideographs solely according to word order and the placing of particles. Many of its words may be used as nouns, adjectives, or verbs. The sentences are loosely strung together in a sequence determined more by emotional content than grammatical rule. The Chinese word is not an abstract sign representing a clearly delimited concept as it is in English. Instead, it is a trigger that releases an indeterminate complex of pictorial images and emotions when seen. The intent is less to express a rational idea than to influence the listener. The manner of blending these emotional cues to yield meaning is more often the result of the sota of the listener than that of the speaker.

No effort to imitate the Chinese sentence in English can be entirely successful. Still, the emotional force of the Chinese language may be felt, although in a rather insipid way, by reading down the two columns in Table 1, one after the other. As each word is read, it should be replaced by the most graphic mental image you can conjure of what it represents. Somehow we feel different as each column is read.

This effect would be heightened if instead of words, a master artist were to paint a sequence of pictures to replace the words in each column or a master poet were to do the same with a line of poetry. In this example, as in Chinese, meaning resides in the Gestalt produced by an accumulation of individual elements. English, Chinese, and many other languages are fundamentally different in their construction, content, and intent.

Since languages are so fundamentally different, as Americans we must be either extraordinarily lucky or excessively chauvinistic to find that English (which seems so suited for the logical, rational hustle-and-bustle of the banker high in a skyscraper in New York City) can *necessarily* capture the subtlety of the vision that *All is heuristic* better than quiet meditation (which seems so suited for the life of a hermit high in the Himalayas in India). This possible incongruity of a language and an idea is even more strongly felt if you are familiar with the highly intuitive language one Japanese uses when he talks to another Japanese. Japan was isolated for so long and the shared tradition of its inhabitants so strong

TABLE 1 Simulated Chinese Language

Column A	Column B
Madonna	Torture
Child	Guts
Serene	Mutilation
Green	War
Blue	Red
Horizontal	Orange
Deer	Acute
Fawn	Holocaust
Peace	Hitler
Christ	Gas chamber
Hope	Hate
Love	Gun

that it is what is not said and how what is not said is said that conveys meaning between native Japanese speakers. For an outsider, the long pauses, hesitations, and circumlocutions are as excruciating as they are inscrutable.

The multiplicity of radically different languages certainly casts doubt on any one language's ability to capture reality accurately and completely. To answer this concern some scientists have been led to take the drastic step of assuming that more than one reality exists, each created by a different language. To quote the opinion of one linguist, Benjamin Lee Whorf, in *Language, Thought, and Reality:*

> We are thus introduced to a new principle of relativity, which holds that the same physical evidence does not lead all observers to the same picture of the universe, unless their linguistic backgrounds are similar, or can in some way be calibrated.

In the view of some linguists, we live in different worlds, literally.

The Western and Eastern views of reality are certainly different and seem to coincide with the difference in languages. As if by magic, Western philosophers reify laws, causal chains, structure, and order, not to mention such peculiar objects as *truth, essence,* and *being* with only a declarative sentence as a wand. Where is the proof that these concepts really exist apart from the names they are given? As with a golden mountain, a mermaid, and a unicorn, quite possibly none of them is real. The philosopher's writings are often quicker than the understanding and the heuristic underlying the trick remains undetected. Failing to recognize the magic for what it is, the West usually structures Nature in terms of laws, rules, and causal chains and explains it in terms of truth, essence, and being. The sorcerers of the East, unable to evoke many of the spells of the West, are hard at work with their own wands creating their own brand of magic. The Chinese do not seek logic, structure, and order in Nature as much as meaning (Tao) in a world presented uniquely at each moment. Because of the difference in these two linguistic traditions, some linguists have argued that a Western language is better suited for science than poetry and that an Eastern one is better suited for poetry than science. This ability of different languages to create different objects like golden mountains, multiple realities, and the Tao that quite probably do not exist certainly casts a tinge of doubt on language's ability to explain anything faithfully. Instead of stopping here and declaring victory in our quest to show that language is best considered as a heuristic, let us focus on just one reality, the one created by English.

Before we become too self-congratulatory in our extraordinary good fortune in having English to explain our vision, we need to stop and consider our equally good luck in living in the present era, for languages change over time. Most of us can barely understand anything written in the fourteenth-century English of Chaucer, whole sentences by Shakespeare, and the meaning of crucial words in the Bible. The following lines from the *Canterbury Tales*, "Whan that Aprille with his shoures sote/The droghte of Marche hath perced to the rote," today seem badly in need of a good editor, and essays have been written on the meaning of Shakespearean phrases such as "Get thee to a nunnery." As to the Bible, we can only puzzle at the historical flip-flop in the words *ghost* and *spirit*. Assuming that at all times all languages have created a different, but correct, reality sufficient for our present needs would seem to make us accept that the ancient caveman had an adequate language to express and explain everything, including *All is heuristic*, when he uttered his first grunt. Before the imperfect subjunctive evolved, I am quite certain that a mother nuzzled her child. Because languages change over time, an even greater tinge of doubt therefore exists to suggest that language is best considered as a heuristic than our earlier argument that all languages are not isomorphic. Unfortunately even with these new laurels in hand, we cannot stop here for we have agreed to consider two additional reasons for doubt.

Now we seem to be led to assume not only that either we were blessed by speaking the only true language or that all radically different languages are equally true, but also the further proposition that the language we speak at the present moment just happens to be adequate to the task as it changes over time. Our extreme good fortune in having available not only English but also twentieth-century English still does not entirely solve our problems because we each speak our own brand of twentieth-century English. Our words, concepts, structure, accent, and cadence are a personal signature, a very personal signature indeed. The noted behaviorist B. F. Skinner tells us why this is so.

Before reading further, please turn to page one of the Color Insert. First cover the color marked with an A and then the one marked with a D. Do not compare them directly. Ask yourself if you think the two colors, A and D, are the same. If you think they are the same, what name do you give to the color? If you think they are different, what name do you give to the difference? We will return to your answers later.

Communication is behavior; it is something an organism does. If we say that a person is communicating, we should be able to look at the individual and observe him or his actively doing something. This behavior might be talking, explaining, drawing, or describing, but it is doing something. Although using language is behavior, it is not one single, simple, isolated behavior. Instead it is a large assortment of activities technically called a *repertoire of behaviors* that are interdependent and interconnected in complex ways.

Modern learning theory, called behaviorism, explains how a repertoire of behavior is established and maintained in terms of one deceptively simple principle called Thorndike's Law of Effect. This law states: "Behavior is modified by its consequences."

The Law of Effect means that when an organism exhibits some behavior, the likelihood or probability that the organism will repeat this behavior goes up if the behavior is followed by something the organism enjoys (called a reinforcer). On the other hand, it goes down if the behavior is followed by something the organism does not enjoy. Technically, the process is known as *operant conditioning*. Thorndike's law tells parents to do what they have always known instinctively: Praise our children if they do something right and do not reward them if they do something wrong. A large body of experimental evidence demonstrates the accuracy of reinforcement theory in explaining behavior. For example, a pigeon can be taught to distinguish between a red circle and a green one by rewarding it with grain for pecking the appropriate circle, and we see examples of Thorndike's Law of Effect in the compulsive human behavior at the slot machines in Las Vegas. This law has been confirmed and reconfirmed in numerous research studies across the animal spectrum from the lowly worms called *Planaria* to *Homo sapiens* and is no longer considered subject to dispute.

Skinner explains verbal behavior or the use of language by insisting that the words and concepts used in language depend on (expressed more technically, *are contingent on*) the past rewards (again more technically, *past reinforcement*) when the concepts are used. A behaviorist would, therefore, say that a concept we use today depends on its past contingencies of reinforcement. As laypersons we might say that we distinguish one concept from another because the difference between the two matters, or better, because it mattered in the past. The seven words for snow exist in Eskimo because different kinds of snow matter to the Eskimos. In a famous 1922 documentary called *Nanook of the North*, the life of a man in northern Canada was described, including his dependency on a knowledge of the climate to construct his igloo and catch the food he needed each day. If he could not complete his igloo, he would freeze to death; if he could not obtain food through a hole in the ice, he would starve. For either, Nanook needed a very accurate knowledge of the different kinds of snow. The documentary is all the more poignant because two years after the film was made, Nanook failed at these tasks and did indeed starve to death. Here in Austin, Texas, snow is white, wet, and falls from the sky only once every seven years or so. Should we wonder that the various concepts for snow need more elaboration in the land of Nanook than in Texas, where the best my children have been able to do is build a two-foot-high snowman using all of the snow on our front lawn?

Recent results with other primates seem to support the notion that verbal behavior is created and maintained by reinforcement. In a famous experiment, a gorilla named Koko was taught to use one thousand signs from American Sign Language to get food and affection, and more recently an orangutan named Azy was taught to select symbols in ways characteristic of higher language skills to obtain food. According to Skinner, in language, a word must have mattered to exist.

Seldom do we learn a concept suddenly. Instead it is refined slowly over time as it is rewarded in a process called *differentiation*. Differentiation (or shaping) refers to the strategy of building up a complex repertoire of behavior by successive approximations. We have a problem establishing extremely complex repertoires because we would have to wait a very long time for the organism to exhibit the desired behavior spontaneously so we could reinforce it. The alternative is to shape the desired behavior slowly, one step at a time, by reinforcing successive approximations to the final behavior. An example from animal studies will explain this psychological principle.

When a behaviorist wants to teach a pigeon to play basketball using a Ping-Pong ball and a miniature basket, he does not wait for the pigeon to exhibit the complete behavior of flipping the ball into the basket and then reinforce it. Instead, the final, complex behavior is built up by successive approximations. The reward is first given if the pigeon just approaches the Ping-Pong ball. Then it is withheld until the pigeon both approaches the Ping-Pong ball and pecks it. Behavior is built up in this way until the pigeon learns to flip the Ping-Pong ball into the basket. This is precisely what is going on as a child learns a language. Thus, a small child will sometimes call a rabbit a dog, much to the delight of his parents, because he has not yet been reinforced for the appropriate characteristics to distinguish the two. For him, a dog is quite possibly a small, four-legged, furry thing and the size of the ears are yet unnoticed. The concept of a dog, as an adult knows it, emerges over time through differentiation.

Some psychologists have gone so far as to argue that we would not even be able to perceive differences in our world unless we have been conditioned to do so. In this view, the pigeon could not even see the difference in red and green, the child could not even see the difference in a rabbit and a dog, and the Texan could not even see the difference in seven kinds of snow, until the differences were shaped by reinforcement.

Please consider figures A and D of the Color Insert. The two colors are indeed different. Similar samples were prepared in a large format by an art professor at The University of Texas.* He then calibrated them so that 80–90 percent of his art students could tell a difference between the two and name the one at A as pure Cerulean Blue and the one at D as having a very small added mixture of the Raw Sienna that you find marked with a B in the color plate. Over several years, an informal test was performed in a senior class of engineers and nonengineering honor students, none of whom had studied art, to see if they could see the difference in the two colors and correctly characterize it. The results were no better than chance. Only about one-half claimed to see a difference, and none could adequately describe the difference if they saw one. Although admittedly the experiments were not scientific with appropriate controls and sample size, the results were surprising. Now ask yourself if you were able to see and explain the difference. This experiment with color is analogous to the more familiar one with sound. It seeks the equivalent of a person who has perfect pitch. As an additional example of the difference in the perception of color, I believe we all have been with a friend choosing a scarf or tie and have seen colors differently than he has. One of you might have rejected the item because it had a slight tinge of green and did not go with the ensemble, while the other

---

*Art professor Vince Mariani, circa 1985. An effort has been made to reproduce faithfully the exact colors mixed by Dr. Mariani. A slight difference in the colors given here is to be expected and may account for any difficulty you may have had in distinguishing between them.

saw no green at all. We are all aware that some with normal vision can see colors better than others and that a person can be trained to make finer and finer distinctions in the colors he sees. Knowledge of, familiarity with, and reinforcement of the difference in colors undoubtedly affect whether or not we are able to distinguish between them. Some psychologists have even gone so far as to argue that we would not recognize that we lived in a Technicolor world unless colors had mattered to our species in the past. In a later section we will consider the heuristic nature of perception and review this experiment. For now, our interest is only in language, and we simply ask, *How can you give a name to a color you cannot see?*

Even if language were theoretically able to explain reality satisfactorily, perhaps we are not yet able to differentiate the concepts necessary to explain a given idea. We might find ourselves in an analogous position to the nuclear engineer we met earlier who was unable to conceive, much less design, a nuclear reactor until Einstein's heuristic, $E = mc^2$, was known. A person breaking into our conversation at this point would certainly not understand the concept we have identified as a sota and would quite possibly have trouble with the word *heuristic* that we are subtly redefining by successive approximations in the course of this book. In fact, at this point, I suspect that the word *truth* still means something different to you and to me. Our personal languages are simply different. The language each of us speaks may fit our own concepts like a glove, but we all wear a different style of gloves.

We have come a long way in our informal analysis of language. No one can deny that languages are different; no one can deny that languages change over time; no one can deny that we each have a personal version of language. Now let us ask if our personal version of English is always up to the task of explanation.

If we can find at least one situation where language is unable to explain what we know to be true, then we will need a decision rule (a heuristic) to decide if it is adequate in any specific future situation, and the heuristic nature of language is established. Everyone, I submit, has experienced directly, personally, and intimately at least one moment of minor ecstasy when his words were inadequate to express his feelings. Of course, I have no way of knowing when and what that moment was in your life. Whether it was your first view of the Swiss Alps, the Grand Canyon, or the Atlantic Ocean; whether it was your private response to a particular piece of music, work of art, or line of poetry; whether it was your feelings at the death of a loved one, during an act of love, or when your first child was born, I am quite sure that you, and every human, has experienced a fleeting moment when words deserted you. My claim rests on that one Bergsonian moment when the sota defining your language was insufficient for explanation, and before which all theory must bow.

But what if you deny that such a moment has ever been yours? A sad situation to be sure, but perhaps you will accept the word of the many poets, artists, Eastern philosophers, and theologians who tell of the fickleness of their muses and nod in agreement with Dante when he cries out, "My vision is greater than our speech." Eugene S. Ferguson in *Technology and Change* by Burke and Eakin reminds us that Albert Einstein claimed that he rarely thought in words at all and that his visual and "muscular" images had to be "laboriously" translated into conventional language and symbols. Other modern scientists have thrown up their arms in despair and admitted that our language is inadequate to describe the atomic and subatomic world. Physicists who endorse the Copenhagen Interpretation of Quantum Mechanics have been forced by their own findings to acknowledge that knowledge of reality lies beyond the capabilities of rational thought. Whether the moment was personally or vicariously experienced, no abstract theory of what language can do prevails over what it once could not do.

Still there will be those who, insensitive to the world around them and unswayed by the testimony of others, will insist on a demonstration that the sota called language is at times inadequate. Fortunately, such a demonstration exists. Ironically, the premise that language is able to explain everything cannot itself be adequately explained. Turning a concept, in this case, *explanation*, back on itself is obviously a peculiar tactic, but it is one dear to the heart of the skeptic, who throughout history has been able to outflank other even more formidable opponents with this maneuver. To show that the adequacy of language in all situations cannot be demonstrated and to introduce the power of the skeptic whose acquaintance we will need later, let's watch him spar with the proposition "Language is able to explain everything."

The effort to establish this proposition rationally fails for one of three reasons: It degenerates into an infinite regress, it requires an unwarranted assumption, or it reduces to a vicious circle. In other words, it succumbs to a slightly expanded form of one of the most powerful weapons of the ancient skeptic, the *diallelos*. First, we ask for an explanation of the claim that language can explain everything. We want to know what this phrase means. But in order to understand the offered explanation, we will need a second explanation of the first one. But then again, we will need a third explanation for the second one and so on, ad infinitum. Completing this infinite program is clearly impossible and the effort to explain explanation on this basis fails. To avoid an infinite regress, we might assume a starting point in the form of an explanation that needs no explanation; that is, we might base our argument on an unwarranted assumption. This second prong of the *diallelos* is clearly unsatisfactory. Actually, in this specific use of the *diallelos*, a stronger condition holds: If you cannot explain this starting point (whether or not it needs an explanation), a counterexample to the original claim that language can explain everything has been found. If on the other hand you can explain why an explanation is not needed, how do you explain that explanation? We now find ourselves, once again, back in an infinite regress. A third possibility is to go outside the notion of explanation for its validation. For example, we could simply ask for an adequate proof (or sign or whatever) that an initial explanation is correct and then use it for further work. But proof itself is a matter of inquiry, and it, too, succumbs to the *diallelos*. (How are we to know that the proof is valid? If by another proof, we have an infinite regress; if by assumption, we have an unwarranted starting point; and so on.) To avoid these known traps, we might be satisfied with an adequate explanation of why the proof should be taken as valid. This, of course, creates a vicious circle: an explanation requiring a proof; a proof in turn requiring an explanation. Like the boxer pummeled by a jab, a cross, and a hook, the claim that language can explain everything falls to the infinite regress, the unwarranted assumption, or the vicious circle. By the way, if you do not believe that the *diallelos* always wins by a knockout, try explaining why not. If you cannot, and many have tried and failed, then the original claim that language can explain everything falls to a one-two punch)

Of necessity, this discussion of language has been limited to four rather simple considerations instead of entering into the endless, often subtle, arguments among recent philosophers over the true nature of language. I believe we can sense the heuristic nature of language directly for ourselves. Just ask yourself these questions:

1. Can this book be translated into another language with absolute faithfulness?
2. Would you read the exact same book if you were to read this book again?
3. Do all people read the exact same book as they scan these words?
4. Are you absolutely certain you understand my meaning?

Given the large number and wide diversity in languages; given the change in languages from guttural grunt to the sophistication of the imperfect subjunctive; given the personal, private nature of the language we each speak; and given that one time that language has failed each of us, surely it is not too much to admit that language is plausible, helpful, based on experience, useful, but perhaps unjustified, incapable of justification, and fallible. Based on our intuition and this brief discussion, I believe it is safe to conclude that we should replace *language* as an absolute with the

HEURISTIC: Language.

If you still have any doubt about this matter, feel free to stop and review the technical literature referred to earlier. The issue of whether language is a heuristic or not is far too important to pass by unresolved.

Just as we earlier agreed to consider any argument based on induction (Bacon's *Novum Organum*) as a heuristic argument, from this point on, any argument made in any language will also be considered as a heuristic argument, including this present argument, of course.

This digression to establish language as a heuristic was desirable to suggest the problem of using a small subset of all heuristics (that is, one specific language, used by one specific person, at one specific time) to explain the heuristic. When we restrict ourselves to language to paint a world in which everything is a heuristic, we are limited to one color, far too few to do justice to our subject.

With this caveat in mind, language will now be used to point beyond itself and explain the meaning of the phrase *All is heuristic*. But just as "the finger is no longer needed once it has indicated the moon," language will no longer be needed once the nature of the heuristic is felt. If you do not keep this simple fact before you, you will not try to remove your dependency on language as a crutch, and the kōan that began this section will defeat you.

Just identifying language as a heuristic is not sufficient for our purposes; we must constantly keep this fact before us. An ancient Arabic fable cautions against ever letting a camel's nose get under the tent. If you do, according to the fable, soon the nose will be followed by the head, neck, and body of the camel until at last the entire camel is inside, and the tent collapses. This fable is instructive with respect to using language to explain anything. If we forget that language is a heuristic even for a second, like the Arab's tent, our philosophical system will collapse. Let me employ a simple orthographic device to serve as a constant reminder of the ease with which the camel's nose can get under the tent when we forget that language is a heuristic. From time to time the second letter of a word will be underscored to emphasize the word's heuristic nature. For example, if I write "*All is heuristic* is true," you may or may not believe me, but in any case the discussion revolves around arguments for one side or the other of the proposition. Strangely this argument contains the seeds of its own demise. A prior argument should concern whether *true* should or should not be written as *true*; that is, whether or not *true* is a heuristic. Writing "*All is heuristic* is true" reminds us of that point. This departure from conventional English may be objectionable to some, but it will show the desperate lengths to which I am willing to go to keep the heuristic continually before us.

As you have no doubt noticed, from the beginning of Part IV, the central claim of this chapter has been written *All is heuristic*. Underscoring the second letter of each word is a frank admission that *all, is,* and *heuristic* are being used as heuristics to convey my meaning with full recognition that they do not necessarily have an absolute significance. After we have made the phrase *All is heuristic* compelling, we will return to each of these words for a closer analysis of its heuristic meaning.

Actually, of course, not only individual words, but also all aspects of communication are heuristics. This includes its grammar, logical connectives, and so forth. In fact, entire ways of reasoning and entire arguments are susceptible to our ornery camel. If we are not extremely alert, we can be tricked into letting the camel's nose under the tent and accepting an argument as convincing when in fact it is only a heuristic itself. To prove my point, reconsider the *diallelos* mentioned earlier. This was the argument that all explanations must succumb to an unwarranted assumption, an infinite regress, or a vicious circle. These three options seem undeniably to cover all possibilities. Where is our guarantee that a new prong to the *diallelos* will not be found in the future? All three of the ones mentioned here have not been around since the *diallelos'* first appearance. As it stands, it is so persuasive that we can easily overlook that it, too, may only be a heuristic. We should have written *diallelos*. Too convincing an argument is not convincing because its conviction depends too heavily on its being convincing.

For practical reasons, underscoring the second letter in only the worst offenders such as truth, absolute, clear and distinct, reality, and similar words must suffice to remind us that everything is a heuristic. This procedure effectively creates a nonassertive language powerful enough to write the skeptic's dangerous sentiment "*I doubt everything*" in such a way as to avoid the implication that he has found one thing beyond doubt (to wit, the fact that he doubts). Writing "*I doubt everything*" avoids the dogmatist's supposed fatal attack to the skeptic's jugular.

From the very beginning of this discussion I have been keeping a running list of important concepts whose absolute nature I was required to assume to make progress. In each instance there was sufficient dispute among experts to consider each as a heuristic. With our newly acquired convention, I really should go back and underscore the second letter in each instance of the words *classification, change, definition, set, ethics, induction, time,* and *language* in the early pages of this book.

Underscoring the second letter of words as a reminder of their heuristic natures has additional advantages. For one, it makes it hard to quote a sentence from this discussion out of context by forcing any quotation to carry its implicit heuristic connotation along with it. When at last I claim to have found "t<u>h</u>e method," by underscoring the second letter in *the* there can be no confusion as to what I mean. Underscoring letters in words in other books on philosophy as you read them is also a worthwhile and revealing exercise, for doing so will make you immune to the prose of the persuasive dogmatist. Somehow *The e<u>s</u>sence i<u>s</u> to be p<u>e</u>rceived; I t<u>h</u>ink, t<u>h</u>erefore I a<u>m</u>;* and *A<u>c</u>t o<u>n</u>ly on that maxim through which you c<u>a</u>n at the same time w<u>i</u>ll that it should b<u>e</u>come a <u>u</u>niversal l<u>a</u>w* seem far less convincing than the originals. Having used this trick successfully to control the invasion of camels in others' work, to be fair, from now on I will use an occasional underscore to bludgeon any k<u>n</u>ows that tries to appear in mine.

COMPELLING BELIEF IN KŌAN

The second objective of this part is to compel belief that *Al<u>l</u> i<u>s</u> h<u>e</u>uristic.* Contrary to what is often thought in the West, rational argument is not the only legitimate method for compelling belief. Some find that behavior modification (and its extreme form, brainwashing), the sudden physical blow administered to monks in some monasteries to force enlightenment, and, of course, prayer and meditation are equally effective ways to change opinion. One additional approach deserves special attention because it bears a slight relationship to the method I intend to use to compel belief that *Al<u>l</u> i<u>s</u> h<u>e</u>uristic.* In some Buddhist training, kōans lay traps for the unwary monk who fails to remember the provisional nature of all human concepts and confuses his world with an assumed real one. The novice is asked, for example, the kōan, "What is the sound of one hand clapping?" or "What did your face look like before you were born?" If we may believe the Zen master, in the frustration of trying to answer these kōans is born belief and enlightenment. Whether by rational argument, reinforcement theory, or a blow from a stick, convincing others that a particular set of heuristics or sota is the only correct one has been a favorite pastime of all people in all cultures. Only the techniques to do so are different.

Basic Approach to Compel Belief

All of the strategies just mentioned are useless for compelling belief that everything is a heuristic, for obvious reasons. Because of its popularity in the West, rational argument requires special consideration to show its inadequacy and to compare it with the approach that will be used here.

Philosophers making rational arguments to persuade us of their view of the world remind me of eccentric old women at a knitting bee. The philosopher's arguments consist of simple steps that follow one another linearly like a piece of yarn that twists, loops, knits, and purls to become a sweater to protect us from the cold, unknown world. One old lady rocks in the world of her fancy, oblivious to others around her except for an occasional pause to criticize the work of her immediate neighbor. She drops stitches as she knits, and, on closer inspection, her initial knot is always found to be unsecured. As a result, all too quickly her sweater begins to unravel. Another lady sitting close by, concerned about the dropped stitches of the first, begins her own sweater using a different pattern but also fails to secure the first step. Others nearby knit with yarn of so many colors and with such a complex pattern that it is nearly impossible to tell where the work is flawed, and a few knit so elaborately that no one cares to examine it closely. In every sewing circle, we find the individual who spends all of her time seeking a pattern and never knits, the one who questions whether we would not be better off left naked, and the one who refuses to participate convinced that if a sweater is needed, it will be provided. Like a knitted garment, conventional arguments depend on an initial starting point that is assumed to be true and an agreed-upon system of logic to pass from one step to the next. In the world of the heuristic, neither may be taken for granted.

My argument in support of the claim that *All is heuristic* will resemble less a knitted garment than a complex tapestry woven from independent, dissimilar strands. Each strand is a heuristic that interlocks with others to produce an integrated whole. If skillfully implemented, this strategy should be as prized as the philosopher's argument for its overall effect, intricacy, and ultimate strength. If a thread snaps in my tapestry, another one may easily take its place. If an area becomes worn or outdated, another patch may be just as easily interwoven into it. At each moment my argument is as complex, relevant, and convincing as the contemporary sota will allow, and since it is woven without selvage, it is intended to blend imperceptibly into your sota when completed. Because my system does not allow me to define, to prove, or to argue, I now plan to cajole, to coax, and to wheedle you into accepting my thesis by weaving such a tapestry. Its motif *All is heuristic* will only achieve its full effect when we have finished.

I weave in the netherland peopled by specters in the form of heuristics. This is perhaps an illusory world to be sure, but it is the one in which I live and where I must make my claim compelling. If explanation in this world is hard, proof is theoretically impossible. To explain what my kōan means, I only ask you to pretend, to imagine, to assume that the concrete concepts that fill your world have become the shades that rode in mine. We must pretend that time, space, and causality are heuristics. We must imagine a world in which truth, being, doubt, and, ultimately, even the heuristic itself have become questionable. While this tactic may be satisfactory to convince, it is not powerful enough to prove. You want proof that what I say is true; in my world I can only offer you proof. You demand evidence, confirmation, or verification; instead I propose evidence, confirmation, and verification. Living in your world and as yet unconvinced, you will find my method disingenuous. It is not that I have no way to convince, to persuade, to compel you to believe; it is that our relationship to this method is different. If you voice acceptance of any strategy I use to support my claim, with the same breath you reject the claim itself. In essence, you can only accept my claim by rejecting all demonstrations of it. With imagination too insipid to transport you to my world, my method to compel belief is to wheedle you into Charon's boat and make you drink of the River Lethe.

Since I hold that my argument must necessarily be unconventional, I caution against seeking order, unity, and continuity in what follows. Instead it should be treated as a pleasant but essentially undirected and arbitrary excursion along one strand of heuristic after the other, a typical discussion as it were. When I gave an early draft of the next section with its enormous number of random, unrelated heuristics to a colleague, he evidently forgot this advice. To quote my overly modest friend:* "When my simple mind encounters a list longer than four

---

*Dr. Dendy Sloan, 1982.

to seven things, I become befuddled." He then suggested that I either provide a detachable study guide or reduce the total number of heuristics treated considerably. His frustration was, of course, precisely the point. I wanted him befuddled. To achieve the effect of the Zen kōan, the large number of heuristics given is essential for frustrating any attempt to keep all of them in mind at the same time, for trying to relate one to the other in a logical way, for building one upon the other, for judging, for endorsing, or for rejecting. To believe my premise, you must come to accept, as Bertrand Russell was at last forced to admit, that order, unity, and continuity are only human inventions—or, as I would prefer to say, only human heuristics.

My Program

With the caution not to seek logic, with the caution not to seek rational argument, with the caution not to seek chronological, historical, or rhetorical order in what follows, I will interlace two sets of heuristics to compel belief that *All is heuristic.* You should, however, consider the following two sets as only a simple model for the much larger number of heuristics that actually exist in each person's sota.

The first, or warp, is a set of rather colorless and uninteresting heuristics chosen to suggest that other concepts, notions, and objects are themselves heuristics. In effect, the warp heuristics bring other concepts into the heuristic fold. Its members are clearly heuristics, often overlapping, uninteresting philosophically, occasionally frivolous, generally applicable, and not really fundamental or essential to the Western worldview. The following set is the foundation, freely admitted to be heuristic, with which my loom is strung:

1. If my necessarily superficial treatment of a complex subject grossly or mischievously misrepresents the current sota of the expert, some critic somewhere will protect your interests and cry foul.

2. If all experts agree on a matter, then we should heavily weigh their opinion.

3. If we are unable to find a single example of a concept, we should suspect that it may not exist.

4. If direct quotation from creditable sources suggests caution in accepting the validity of a concept, exercise caution yourself.

5. If qualified experts in the *same* field hold widely differing opinions on a matter, then we should suspend judgment.

6. If a concept produces paradoxes, unexplained complexities, or departures from expected results, better consider it a heuristic.

7. If the literature treating a concept is enormous, we should be suspicious that the matter is far from settled and that we are dealing with a heuristic.

8. If experts in widely *different* disciplines agree that a concept is in question, perhaps we should do the same.

9. If experts in widely different disciplines disagree as to whether or not a concept is in question, perhaps we should be cautious.

10. Sincerity of belief and inability to disbelieve are poor justifications for claiming that a belief is true.
11. Beware the certainty of the captured expert.

This warp may be recognized because each heuristic strand is a complete sentence. As we read down this first list, each entry seems plausible, but fallible.

The woof is a different matter. I have a little list of concepts that will not be missed as absolutes once they are identified as heuristics. They are seldom considered as heuristics, extremely interesting philosophically, very specific, and essential to the Western worldview. My overall program, a very ambitious program at that, is to reduce each of them to a heuristic in a few paragraphs. Of the many possible colors, sizes of thread, and materials from which I could choose to weave into my warp, I hope the following are sufficient to make my point:

1. Arithmetic
2. Mathematics
3. Deduction
4. Certain
5. Position
6. Logic
7. Truth
8. Progress
9. Causality
10. Consciousness
11. Physical reality
12. Science
13. Perception
14. Argument

When we first considered the words *classification, change, definition, set, ethics, induction, time,* and *language,* we could not tarry overly long to consider their heuristic nature. Were time to permit, they could be added with profit to the beginning of this second list.

The woof may be recognized because each heuristic is a simple word or two and represents a specific concept. As we *now* read down this list, each entry seems absolute, universal, and beyond dispute. Periodically these two lists will be reexamined to evaluate our progress.

A cruel trade-off between specialization and generalization complicates the intellectual's world. The expert is both the least qualified person to accept the claim that everything is a heuristic and the single most important individual for encouraging others to do so. We encountered this problem earlier when we considered the problem of specialization in engineering.* An engineer must spend a lifetime of study to be at the state of the art in, say, the strength of airplane wings, which leaves no time for him to be knowledgeable in the overall design

---

*See Figure 24.

of airplanes. In a similar fashion, a person who has spent a lifetime studying one isolated area of knowledge and can truly be considered an expert in it is not particularly well prepared to make a grand generalization that encompasses all areas of knowledge. To convince that everything is a heuristic, we are constantly stymied by this subtle trade-off between specialization and generalization. We will have more to say about this problem later.

For this part of our discussion to be successful, I will need to include a large enough list to befuddle and thwart any effort to use logic and rational thought, a sufficient number of items to let the heuristic _induction_ work its magic, but still not include so many as to bore. No one, certainly not myself, could ever hope to be an expert in all of the items on just the second, short list. This would require an intimate knowledge of Gödel's Proof, quantum mechanics, the Heisenberg Uncertainty Principle, Hume's attack on causality, Bell's Inequality, and Pyrrhonism as an absolute minimum. Each of these topics is far too complicated for one person to master without a lifetime of study. If, by chance, a person knowledgeable in all of them were found, I would simply augment my list to include Church's Theorem, the Ergodic Assumption, the infinite, string theory, the Cosmological Assumption, and any of a large number of other concepts that I have found to be heuristics in over thirty years of searching. I am confident that eventually he would reach the limits of his ability to bring order, unity, and continuity to his analysis.

On the other hand, an expert in each of these individual areas on our list is your best safeguard that what is reported here is true. How can I avoid overwhelming the reader who has never stopped to consider each of these matters deeply or underwhelming the reader who has done so? As you may no doubt have guessed by now, I call on the first warp

HEURISTIC: If my necessarily superficial treatment of a complex subject grossly or mischievously misrepresents the current sota of the expert, some critic somewhere will protect your interests and cry foul.

This is clearly a heuristic. Experts have mental lapses; experts have private agendas; experts can be envious, devious, and disingenuous. In fact, an occupational hazard of the expert is to know he is one and to challenge anyone who takes a contrary view or to criticize anyone who does not come up to his level of competency. Still the opinion of an expert who is generous and understanding of the problem I have in explaining a list of complicated concepts in a very limited space is your ultimate safeguard that the short descriptions of each woof item are not grossly or mischievously misrepresented. As I warned in the Introduction, throughout this discussion I have surreptitiously sprinkled "almost quotations" from some of the most famous minds, past and present, to sustain my claim. I trust that none of them abuses the good nature of those to whom I stand in debt. _For my program to be successful, what we need is a shallow draught of a large number of Pierian Springs with a local health official to certify the water in each._

Before we begin to weave the two lists together, we should take note of two important problems of scale that govern our task. In a widely imitated essay, the Dutch writer Kees Boeke invites us to consider the vastness of the known universe by considering a cosmic view of a man and woman lying on a blanket from a distance that successively increases by orders of magnitude and then to do the same for decreasing orders of magnitude. Recent interpretations indicate that the known universe extends over forty powers of ten. That is 10^{40} or 1 followed by forty zeros, thus:

10,000,000,000,000,000,000,000,000,000,000,000,000,000.

The distance in which we are imbedded is truly beyond comprehension. Focus now on time. If the height of the Empire State Building in New York City represents the age of the earth (2.6×10^5 cm or 4.5 billion years), the age of man would equal the thickness of a nickel. On that scale, recorded history would be represented by only a scratch on the nickel. The point is that you and I, as individuals, represent a unimaginably small part of space or time. The notion that any concept conceived by the human intellect has ultimate standing is unabashedly egocentric. The extension from knowledge of earth today to the cosmos of eternity is an extrapolation from known data that boggles the mind of any honest, practicing engineer.

Even an outline of my argument is complex. The warp items on the first list will be considered in turn and applied to one or more of the items on the second list of woof heuristics. In other words, my demonstration is complicated because each warp heuristic interacts with more than one woof string, and each woof heuristic interacts with more than one warp string. This mutual interaction of heuristics, modeled here by two arbitrarily chosen groups of heuristics, actually applies to all heuristics in each person's sota. Meaning resides in the Gestalt produced by the overwhelming accumulation of individual heuristic interactions, not in any particular choice of individual concepts. To achieve the maximum effect, I suggest that you read from here through page 190 as quickly as you can without backtracking—at one sitting, if at all possible.

In my world I cannot knit; please join me as we weave.

Weaving a Tapestry

All is heuristic. I have now said it thrice; what I tell you three times is true. Would that this heuristic of Lewis Carroll for achieving certainty always guaranteed an answer, and I could leave it at that. Unfortunately it does not, and so we must now consider the next heuristic that makes up the warp of our analysis, the

HEURISTIC: If all experts agree on a matter, then we should heavily weigh their opinion.

Of course, this heuristic does not guarantee a solution either. Experts have been wrong in the past and will certainly be wrong in the future. Our job, however, is not to produce certainty, but the best understanding of our world. For one who has not studied a matter himself, the opinion of experts should not be lightly discounted, or to use a Latin expression, *experto credite.**

Experts may agree either that a particular concept is in doubt or that it is certain. In the first case, we have one more item to count toward everything being heuristic. Given the enormous number of concepts whose status could be ex-

---

*"Believe one who has had experience."

amined, agreement that one more item is in question contributes very little to our central claim unless it is truly unexpected. In the second case, agreement among experts that a concept is certain produces a clear counterexample to this claim and is more serious to our cause. The current status of these two possibilities lends support to the view that everything is heuristic. First, let us consider a concept that unexpectedly all experts agree is in doubt.

Arithmetic as Arithmetic

Although the first warp heuristic seems beyond dispute as a good heuristic, one surprising, little known, and, to the uninitiated, unbelievable example of a concept that all competent experts in the field now feel is tinged with doubt may make you wish this were not so. It is the woof

HEURISTIC: Arithmetic.

Arithmetic, that is, the system that gives us $2 + 2 = 4$, is fundamentally flawed. On this all experts who have studied the foundation of mathematics agree. If you do not have a lifetime to spare for the study of number theory, you must accept the silence of the critic as certification that this characterization of arithmetic is accurate. *Experto credite.* Kurt Gödel, an American logician of Austrian birth, originally made this diagnosis in 1931, and the problem lingers still. Arithmetic is so badly flawed that I feel justified in calling it a heuristic.

I admit that I was disconcerted when I first encountered Gödel's Proof. While I was learning chess and studying artificial intelligence in the spring of 1965, my major professor* tossed a little, buff-colored book by Nagel and Newman describing this proof across the desk and suggested that I look it over. My first reading ended at 1:30 A.M.; my second, at 5:30 A.M. I was astonished. Montaigne may have had his Sextus Empiricus and Kant his Hume to usher in their skeptical crises, but I had Gödel to usher in mine. I was so overwhelmed by its implications that I now consider this proof the second most important idea ever conceived by the human mind.[†]

With a bit of serendipity, within the week, I, along with the other students in my artificial intelligence class, was required to program a computer to demonstrate a portion of Gödel's Proof,[‡] and my crisis deepened. Imagine how unnerved an apprentice engineer becomes as he sees, both theoretically and computationally, the certainty of mathematics dissolve before his eyes. Still, I knew I had to complete my engineering homework for the next day, and I badly needed mathematics to do so. Fortunately at the time, the same course in artificial intelligence was developing a peculiar, nonalgorithmic way to program a computer called heuristic programming. I had my answer: arithmetic was a heuristic! Arithmetic might only be a heuristic, but clearly it was a good and very necessary one. All the while I could not help wondering—"If arithmetic is in

---

*Dr. Kent Hansen.

[†]Everyone it seems has his choice for first on the list and will fight to the death for this idea. For some it is the Bible, for others the Talmud. To my knowledge no one fights for second place so I'm secure in achieving this rank and can hopefully compel you to agree.

[‡]In a course in the computer language, LISP, by Profs. D. Bobrow and M. Minsky.

doubt, what is not?" Slowly, the phrase *All is heuristic* was born, and this discussion some thirty-eight years later is the result.*

Gödel's Proof (actually, two proofs) is difficult, very difficult. According to my buff-colored book, forty-six preliminary definitions, together with several important theorems, must be mastered before the main results are reached. Because of the complexity of Gödel's Proof, the best most articles intended for the non-mathematician can do is struggle to give a general flavor of the proof. I intend to follow this same approach, but being an engineer, I shall use successive approximations to the final proof. Feel free to stop and skip to the summary at the end of this analysis as soon as you are satisfied with your understanding of the matter and are willing to trust the view of the experts. Essentially all of us will have to do this eventually, since none of us is probably equipped to see the proof through to the end.

First Approximation

Basically Gödel's Proof is a mathematical statement that asserts of itself that it cannot be demonstrated. That is, this mathematical statement shows that it cannot be derived from the primitive propositions and definitions of the language of which it is a part by applying the rules of inference of that language. More succinctly, it takes the form of a proof that arithmetic is either incomplete or, perhaps worse, inconsistent. In informal terms, incompleteness means that relevant truths within arithmetic cannot be derived from the initial axioms; inconsistency means that a falsehood or contradiction can be derived. In the first case we are dealing with a heuristic; in the second case, even more clearly so. If arithmetic is incomplete, how can we ever be sure that one of those truths we cannot prove is not absolutely necessary for our understanding of a matter under consideration, except heuristically? If arithmetic is inconsistent and we can derive a fact as both true and false, either avenue we pick is clearly a heuristic, and, if I may say so, quite possibly a very bad heuristic at that. If you are satisfied with this characterization of Gödel's Proof, feel free to say the magic words *experto credite* and skip to the conclusions on page 143.

Second Approximation

Some may wish a deeper insight into this proof that is so pivotal to our work. This second approximation and the ones to follow will examine the two potential problems of arithmetic predicted by Gödel's Proof by creating a model of each that is easier to understand. *Inconsistency* will be modeled in language, and *incompleteness* will be modeled in a Texas legend concerning the Ghost Riders in the Sky. These two models skirt the difficult aspects of the proof and allow us to focus on the kernel of the matter. We will leave it for the experts to map each model into arithmetic and to show that the characteristics we study do indeed apply.

In this approximation and the next, we attack the problem of inconsistency. Let us begin with definitions of two important concepts, a *metalanguage* and an *antinomy*, to help clarify the comments that were made in our first approxima-

---

*In a 1982 version of this book for classroom use I gave seventeen years. Now in 2003, it has been thirty-eight years and I am even more convinced that this is true.

tion. We will use a series of exhibits to explain what each means. (We depend heavily on the work of Tarski and Quine in the analysis that follows.)

A metalanguage is any language that talks about another language. Our world contains many objects such as chalk, balls, people, cars, and so forth, and we can talk about them. Exhibit 1 is an example of a sentence in English.

CHALK IS WHITE.

Exhibit 1

In this English sentence we are commenting on one of the objects in our world, specifically the one that goes by the name of chalk. It has a subject and a verb. It is a complete sentence. To make sure that we can always recognize a sentence, word, or phrase in English when it appears in one of our exhibits, we have used bold italics. With this notation, you can always locate the part of each sentence in our examples that is in English.

The language *English* is, itself, made up of objects in our world that we might want to discuss such as nouns, verbs, adjectives, adjectival phrases, sentences, sentence fragments, and so forth. To talk about them, however, we will need what is technically called the metalanguage to English. Exhibit 2 gives an example of the metalanguage to English.

CHALK HAS FIVE LETTERS.

Exhibit 2

In this exhibit we are talking about an object that makes up the language *English*, specifically the noun *chalk*. We have made a definite assertion about this object: It has five letters. We recognize the word in bold italics as being an element of English and will take the plain text as being the metalanguage to English. Of course, this metalanguage must also contain the elements of the language upon which it is commenting so it can talk about them. Although our use of bold italics to distinguish between English and its metalanguage is not the standard notational convention in the philosophical discipline called logic, it is a standard convention in English style books such as the *Chicago Manual of Style* and is familiar to most of us.

We can also talk about a complete sentence in English in its metalanguage, as in Exhibit 3.

ALL IS HEURISTIC IS TRUE.

Exhibit 3

Ignore our previously discussed convention of underlining the second letter of troublesome words, for it has no relevance to the matter at hand. In this exhibit we have no difficulty picking out the part in English for, once again, it is in bold italics. The entire example is in the metalanguage to English because we are talking about a sentence in another language.

Even at this early stage in our venture, we have omitted several important considerations that would provoke a sidebar in a class in the study of logic. One of these is the difference in an object and its name. In the previous example the word *chalk* is the name of an object and obviously it is not chalk itself. Likewise the English sentence in Exhibit 3 is technically the name of a sentence and must be considered as a unit. But since we seek only the flavor of Gödel's Proof, I am sure that the generous expert in logic will give us some leeway and let us omit some of these considerations that are extremely important to his work.

English also contains what are called sentence fragments; that is, it contains collections of words that do not quite make the grade as a sentence because they are missing an important component such as a subject or a verb. Not too surprisingly we can also talk about them using a metalanguage, as in Exhibit 4.

YIELDS A FALSEHOOD IS A SENTENCE FRAGMENT BECAUSE IT DOES NOT CONTAIN A SUBJECT.

Exhibit 4

Our freshman English teacher cautioned us against using similar sentence fragments in our themes and quite possibly marked us down for using them in our papers. To make him happy, we would have to add a subject to our fragment so we know exactly what it is that *yields a falsehood*.

We can also have a metalanguage to the metalanguage to English. That would be a language that talks about the language that talks about English. You have already seen an example of this situation in this book and it caused us no trouble. It is reproduced in Exhibit 5.

WRITING "*ALL IS HEURISTIC* IS TRUE" REMINDS US OF THAT POINT.

Exhibit 5

In this case we have used quotation marks to embed the sentence in the metalanguage to English within its own metalanguage more in keeping with standard practice. In fact we can have an infinite regress of metalanguages, each commenting on the one below it.

With the definition of a metalanguage in hand, let us see what mischief we can cause as we define an antinomy. An antinomy produces a self-contradiction when we use accepted ways of reasoning. Since we will soon discover that the accepted way of reasoning in Gödel's Proof is the whole of mathematics, deduction, and the scientific enterprise, we can anticipate that we are in for big trouble.

In Exhibit 3 we considered the truth or falsity of a sentence in English. In the spirit of Aristotle's definition of truth in Metaphysics we may write claims of this type in the metalanguage to English, as in Exhibit 6.

SNOW IS WHITE IS TRUE IF AND ONLY IF SNOW IS WHITE.

Exhibit 6

We understand this sentence to mean that the sentence *Snow is white* is true if indeed it is the case that snow is white. Once again, if this were a class in logic, we might stop to ensure that we do not have a vicious circle, that we have not confused the name of a thing with a thing, and so forth. For this approximation, let us accept this metalanguage sentence at face value. It has a subject and a verb. It is a complete sentence. We understand that it requires that a sentence correspond to the way things really are in order for it to be labeled true.

Many similar forms of reasoning could be explored. Focus on a slightly different form with three examples, as in Exhibit 7.

THE ADJECTIVE *RED* IS TRUE OF A THING IF AND ONLY IF THE THING IS RED.
THE ADJECTIVE *BIG* IS TRUE OF A THING IF AND ONLY IF THE THING IS BIG.

. . .

THE ADJECTIVE *P* IS TRUE OF A THING IF AND ONLY IF THE THING IS P.

Exhibit 7

In the last example, the letter *p* will be taken as a variable and presumably can stand for any adjective or adjectival phrase. Although the expert would have to be concerned about how to make the substitution properly and what font to use to explain what he was doing, this pattern of reasoning seems safe and has a long history. What else could an adjective such as *red* and *big* mean except what is given in Exhibit 7? The antinomy shows its bite, however, when we make one fatal substitution for *p*: the adjectival phrase *not true of itself* and the thing that has this characteristic is the adjectival phrase itself. Then Exhibit 8 follows.

NOT TRUE OF ITSELF IS TRUE OF ITSELF IF AND ONLY IF IT IS NOT TRUE OF ITSELF.

Exhibit 8

Although some quarters might criticize this metalanguage sentence in some aspects, it does present us with, at least, an apparent contradiction based upon our accepted pattern of reasoning. If it is false, it is true; and, if it is true, it is false.

Some of us will recognize this antinomy as a somewhat more sophisticated version of the one with which the class genius we all hated in the sixth grade used to startle us. You remember the antinomy concerning the Cretan Epimenides, who is said to have asserted that "All Cretans were liars." This utterance has perplexed students of philosophy for years for if the statement is true, the fact that a Cretan made it renders it false; and, if it is false, the fact that a Cretan made it renders it true.

These problems revolve around the nature of truth and the use of phrases such as *true of*, and others. This would be an excellent place to begin if we wanted to analyze the paradoxical nature of language according to Tarski that was referred to earlier when we considered language as a heuristic and his conclusion that *true* cannot exist in a universal formal language.

As with most antinomies, Exhibit 8 and its variants have caused a revision in our way of reasoning. Some of the surgery has been draconian. Researchers have proposed that *truth* be eliminated from language, that only certain phrases such as *true of* be eliminated, or that *truth* be indexed depending on the level of the metalanguage used. None of these solutions has carried the day, and the crisis persists today.

This concludes the definitions of a metalanguage and an antinomy and permits us to state a second approximation to Gödel's Proof. Before we do that, several points must be made. First, Gödel did not do his work in English and the metalanguage to English. Instead he used arithmetic and an arithmetical metalanguage to arithmetic. Second, he was not concerned so much with whether or not a statement was true, but with the closely allied notion of whether or not arithmetic could be demonstrated (or was decidable). Instead of truth, he used a characteristic of arithmetic called demonstratibility. Finally, we must once again emphasize that this discussion has neglected many, very subtle, important considerations.

What Gödel did, then, was construct an arithmetical statement in the metalanguage to arithmetic that asserted that arithmetic could be demonstrated if and only if it could not be demonstrated. Perhaps you are now satisfied with this general feeling for Gödel's Proof and are willing to skip to page 143. Throughout history, the incantation *experto credite* has been a powerful restorative.

Third Approximation

Some may want to delve more deeply into the matter. A large number of antinomies such as Exhibit 8 and the one of Epimenides exist. We include only one more because its construction will help those who may later wish to explore Gödel's Proof further. Let us construct a sentence that improves on Exhibit 8 and unequivocally attributes falsity to itself. It is not original with me, of course, but its inclusion here will absorb the literature on logic into our weaving. Begin with the simple sentence containing a variable, as in Exhibit 9.

X YIELDS A FALSEHOOD WHEN APPENDED TO ITSELF WRITTEN IN BOLD ITALICS.

Exhibit 9

Although odd, this seems innocent enough. It has the form of a sentence (lacking a substitution for the variable *x* to make it specific) and with a little thought we understand what it means. The antinomy arises when we make one special choice for *x*, to wit, the fragment *yields a falsehood when appended to itself written in bold italics*. If we do that, we arrive at Quine's antinomy shown in Exhibit 10, and our world becomes curiouser and curiouser.

YIELDS A FALSEHOOD WHEN APPENDED TO ITSELF WRITTEN IN BOLD ITALICS YIELDS A FALSEHOOD WHEN APPENDED TO ITSELF WRITTEN IN BOLD ITALICS.

Exhibit 10

The object in our world that is described by this sentence is precisely this sentence itself, and furthermore it specifically claims that that object (itself) is false. So, once again, if it is false, it is true; and if it is true, it is false. If we had the mathematical background including "forty-six preliminary definitions, and several important theorems" to mirror a form of reasoning quite similar to this in arithmetic and the metalanguage to arithmetic, we would arrive at Gödel's Proof.

Arithmetic is an axiomatic system, and the axioms are well known and understood. Gödel's great achievement was to represent all of the elements of arithmetic in arithmetic itself as a metalanguage. He did not just discuss a certain antinomy concerning the demonstratibility of arithmetic, but he gave a constructive example of one. The result that flows from this example is that arithmetic is either inconsistent or incomplete. All efforts to prove the contrary must involve patterns of reasoning that are more suspect than those used to derive Gödel's Proof. Further, if a person tries to patch up arithmetic by adding additional axioms, he simply creates a new axiomatic system in which a new antinomy can be constructed.

We must not lose sight of the overall goal of this chapter, which is to support the contention that *All is heuristic.* The heuristic nature of arithmetic is the first salvo in our battle to win that beachhead. These results are so startling and unexpected and our prejudice that mathematics is the perfect model of knowledge is so strong that you may want to study the matter further. For those who understandably wish to rush on toward our primary quest, please trust the opinion of all experts in the field and skip to the conclusions on page 143.

Fourth Approximation

The previous two approximations emphasized the problem a complex logic system can have because of *inconsistency* by using self-referencing sentences in English. In the present approximation, we focus on the second difficulty such a system can have, *incompleteness.*

Since this concept is rather difficult to explain and since I'm a native Texan, I'll call on one of my compatriots, the cartoon character Lucky Luke, for help. Lucky Luke is surely one of the most famous cowboys in the world. He is reputed to be so fast with a pistol that he can outdraw his own shadow. He has visited my hometown of Austin, Texas, so you can be assured that I know him well.* You can see Lucky Luke and some of his fellow Texans on page 4 of the Color Insert.

In the bottom figure we see Lucky Luke on the left and his arch rivals, the Dalton Brothers (Joe, William, Jack, and Averell), on the right, along with several additional cowboys of the Old West. Using a cartoon to make an important point is a somewhat odd heuristic in a serious discussion of this nature, but we

---

*Lucky Luke's adventures are chronicled by Morris. See Lucky Luke #13, *Le Juge,* for example.

will be required to keep many new concepts in mind at the same time and a vivid, easily remembered example will help.

Texans are divided into two categories: squeaky clean, polite cowboys and low-down, no-count polecat outlaws. Cowboys always tell the truth. They would never, never lie because their mothers would not like it. Lucky Luke is an example of a cowboy. Unlike cowboys, outlaws always lie. They really do not care what their mothers think because their mothers are even meaner than they are. If a Texan is wearing his hat, you can be sure to which of these two categories he belongs. Cowboys wear white hats; outlaws wear black hats. You can also have confidence that there is no possibility of a mix-up in the hat a Texan is wearing. In Texas, the worst possible sin is to touch another man's hat. Of course, if a Texan is not wearing his hat, we do not know whether he is a good guy or a bad guy and, hence, have no idea whether he is lying or not.

Everyone is undoubtedly aware that in Texas outlaws never die. Instead, when their time has come, they join an innumerable caravan in the heavens immortalized in song as the *Ghost Riders in the Sky*. For their misdeeds outlaws are forced to roundup the "red-eyed devil's herd with brands of fire" across the great range up in the sky. What is less well known is that cowboys join these outlaws in this mysterious realm, but they are given the more enjoyable task of herding Texas Longhorns. On page 4 of the Insert, we see representatives of the cowboys and outlaws (with and without hats) and the timeless roundup in the sky. When it thunders during a heavy rainstorm, a so-called gully washer in Texas, you are actually hearing the Ghost Riders on endless cattle drives across the sky, or so I was told as a child. It is to these Texas cattle drives—drives, for short—we must now turn.

In Figure 31 we see the two groups of Texans: cowboys and outlaws. Each of them has been further subdivided into those for whom we can be absolutely sure that they are telling the truth and those for whom we can be absolutely sure they are lying because we can see the colors of their hats. The figure also shows some for whom we have some doubt to which category they belong because they are bareheaded. We suspect that some of these are cowboys and others are outlaws, but we cannot be sure. In fact, determination of which category the hatless individuals belong to is precisely our most important objective. Finally, any given Texan may have gone on more than one cattle drive.

Let me tell you a few of the characteristics of the Ghost Riders of the Old West that you may not know and then I will ask you a few questions.

COWBOYS　　　　　　　　　　　OUTLAWS

Figure 31 Existence of a Cowboy without a Hat

CHARACTERISTIC 1: All of the cowboys who are wearing white hats have gone on a cattle drive together.

In the square on the far left within the dark border, we see this group. Lucky Luke and Bubba represent them, but many other Ghost Riders should be included. Characteristic 1 tells us that this entire group has participated in a cattle drive together at some time.

CHARACTERISTIC 2: All of the outlaws who are wearing black hats have also gone on a cattle drive together.

In the square on the extreme right within the dark border, we see this group. These are represented by some of the Dalton Brothers.

CHARACTERISTIC 3: Given any specific cattle drive, all the cowboys and outlaws who did not participate in it went on a different drive of their own.

In other words, identify a cattle drive. There will always be a different drive that includes "all but" the members of the first one. We call this second drive the complement of the first one. For example, the drive that complements the one described in characteristic 1 (the cattle drive of all cowboys with hats) would include all of the outlaws whether they were wearing a hat or not and all cowboys who are not wearing their hats. In Figure 31 we have enclosed this second or complementary drive with a dashed line to make it easy to identify. This complement would include the three Daltons, Sonny, and, very importantly as we shall soon see, Jake.

The last characteristic of the Old West mirrors the great discovery of Gödel that leads to the downfall of arithmetic.

CHARACTERISTIC 4 (AFTER GÖDEL): Given any cattle drive, there is at least one cowboy or outlaw who claims that he went on that cattle drive.

Of course he may be lying. In fact, if he is wearing a black hat, we can be absolutely sure he is lying.

Now we need to ask several important questions about a logical system with these four characteristics.

QUESTION 1: Must there always be at least one *cowboy* who is not wearing his hat?

In other words, is there at least one cowboy in the square occupied by Jake in Figure 31? Can this square ever be empty?

By characteristic 4 we know that there must be at least one cowboy who claims to have gone on the cattle drive represented by the dashed line because it tells us that that there is *at least one person* who claims to have gone on every cattle drive. Could this individual possibly be one of the outlaws (whether or not he is wearing a hat)? No. Outlaws always lie and would therefore say that they did not go on any drive that includes them. Therefore there must be at least one cowboy in the dashed area, because we need someone who is telling the truth. The only cowboys we find there are cowboys who are not wearing their hats. The answer to question 1 is, therefore, yes. There must be at least one cowboy who is not wearing his hat.

Surprisingly this is all we need to capture Gödel's notion of incompleteness. The Ghost Riders are a model of any logic system with the four stated characteristics. Informally we can say that the cowboys with hats represent axioms and proven theorems in an axiomatic system and the cowboys without hats represent true statements in the axiomatic system that are true although we cannot prove them within the system. Arithmetic is such a system; arithmetic is incomplete. (To be a tiny bit more precise, the cowboys and outlaws represent strange objects called Gödel Numbers and the notion of truth is really replaced by a mathematical notion called demonstratability. Unfortunately, we must depend upon an expert in logic to straighten out these details.) All that we would have to do to show Gödel's Proof in its entirety would be to exhibit explicitly the correspondence between our model and the elements of arithmetic and show that the four stated characteristics are, indeed, characteristics of arithmetic. Characteristics 1, 2, and 3 turn out to be lengthy, but straightforward to establish for arithmetic. The application of characteristic 4 to arithmetic is a different matter and the approach to it is very clever. We will consider several aspects of this last characteristic in a minute.

Although it is not absolutely necessary to consider a second question to understand Gödel's incompleteness theorem, let's ask it to check our understanding of this analysis. Some find this next question more surprising than the one we have just considered. It will also give us a chance to practice a technique we will need later when we reexamine characteristic 4. Figure 32 will help us understand the second question.

QUESTION 2: Must there be at least one *outlaw* who is not wearing his hat?

In other words, must there always be a Texan in the square where we have shown Sonny? Can this square ever be empty?

Look at the outlaws who are wearing their hats. We know they went on a cattle drive together because of characteristic 2. And, by the famous Gödel characteristic 4, at least one of the Ghost Riders claims to have gone on this cattle drive. Could any of the always-truthful cowboys make that claim? No. Could any of the lying outlaws with hats make that claim? Certainly not. So there must be at least one outlaw who is not wearing a hat, that is, one of the outlaws in the square indicated by Sonny, who made this claim (he was lying, of course, but no matter). Returning to arithmetic, this means that there must be at least one theorem that is absolutely false, but one we cannot prove to be so. Once again, the system is incomplete.

COWBOYS   OUTLAWS

Figure 32 Existence of an Outlaw without a Hat

Careful comparison of this example with the one that preceded it points up an important fact worth remembering. If we need someone who will say (truthfully or not) that he went on a specific cattle drive, we must look for a *cowboy* who is a member of that drive, or for an *outlaw* who was *not* a member of that drive. The outlaw must have gone on the complementary cattle drive as defined in characteristic 3.

Actually, we are making such good progress that it might be fun to carry the analysis one step further. This will give you a chance to glimpse a small bit of the complexity that requires the forty-six preliminary definitions eluded to earlier that are needed for Gödel's Proof.

The cattle drives just considered were really rather simple ones. They included the one consisting of all the cowboys with hats, the one consisting of all the outlaws with hats, and the one that makes up the complement to the first of these. Now we must open our perspective to consider all possible cattle drives. They will include different mixtures of cowboys and outlaws—hatted and bareheaded.

An alternative form of characteristic 4 exists. By showing that it is true and that it implies the original one, we can use it to draw the same conclusions as before.

First consider a few preliminary facts you need to know about our imaginary cattle drives in Texas. Each cattle drive is named after a Texan (one of the cowboys or outlaws) and every cowboy or outlaw has a drive named after him. Each cattle drive has a trail boss, of course. If an individual happens to go on the cattle drive named after him, he acts as trail boss; if he does not, he is just considered a simple hired hand. Figure 33 will make these odd facts clear.

The drive on the left is named after Lucky Luke; the one on the right, after Jim Bob. Note that Lucky Luke is a member of the cattle drive that is named after him, so he is the trail boss. On the other hand, Jim Bob is not a member of the cattle drive that bears his name, so he is a simple hired hand.

Cowboys and outlaws often have sidekicks. We have all heard of the Lone Ranger and Tonto, Gene Autry and Smiley Burnette, and, of course, Red Ryder and Little Beaver. Some cowboys like Roy Rogers, for instance, have had two different sidekicks (Cookie and Gabby). These sidekicks are, themselves, cowboys or outlaws and always tell the truth or always lie as the case may be.

LUCKY LUKE   JIM BOB

Figure 33 Identification of a Trail Boss

Figure 34 Test Cattle Drive and Mystery Texan

Sidekicks are very, very loyal. The important thing to remember about a sidekick is that *he will always say that the Texan with whom he works is a trail boss*. No self-respecting sidekick would ever admit that he accompanies a simple hired hand. (Of course, if we are talking about an outlaw, what he professes is a lie.)

With these preliminaries out of the way, we can state the alternative form of the Gödel condition given earlier. Now for a mouthful that is truly obscure:

CHARACTERISTIC 5 (AFTER GÖDEL, ALTERNATIVE FORM): Focus attention on a specific cattle drive, C. There will always be another cattle drive, D, such that every member of D has at least one sidekick in C, and every nonmember of D has at least one sidekick who is not a member of C.

The problems in understanding this sentence are keeping both cattle drives in our minds at the same time and figuring out where in the world the sidekicks are located. With these two warnings in mind, Figure 34 will offer some help.

In the figure we see two cattle drives: one labeled "test cattle drive" and a fainter one labeled "Mr. X." Our major problem is that we do not know much about this second, mystery drive. We do not know who participated in it. We do know it is named after a Texan because all cattle drives are named after one of the Texans, but we do not know for whom this one is named. Therefore, we have simply called it after a very mysterious cowpoke named Mr. X.

Please put your left forefinger on the first cattle drive, the test drive. It corresponds to C in characteristic 5 that we are trying to understand. We are assured that there will always be a second cattle drive associated with the first. Please put your right forefinger on the cattle drive named after the unknown Texan, Mr. X. This corresponds to cattle drive D. Now we must reverse our thinking and move back from drive D to C (or from our right finger to our left finger). Characteristic 5 tells us that every single Texan in the one pointed to by our right finger, say, Texan #1, will have at least one sidekick back over in the one pointed to by our left finger, say, Texan #2. Once again, point to any Texan in the drive indicated by your right finger, and you are guaranteed to have a corresponding sidekick in the drive indicated with your left finger. Characteristic 5 goes on to tell us an additional important fact. Every Texan who is not in the cattle drive indicated by your right finger (that is, all Texans in the complement to the drive named after Mr. X) will have at least one sidekick who is not in the cattle drive pointed to by your left hand (that is, the complement to the test drive). That is, move your right finger from Texan #1 to Texan #4 and you can always move your finger from #2 to #3. Note that we do not know who these mysterious Texans #1, 2, 3, and 4 are, but we do know that they must exist, and we know their relationship. With these new facts in mind, let us ask an important question:

QUESTION 3: Does this new characteristic 5 imply our original Gödelian characteristic 4?

That is, does this new information about Texas cowboys also mean that "Given any cattle drive, there is at least one cowboy or outlaw who claims that he went on that cattle drive"? If the answer is *yes* (as turns out to be the case), we can make do with characteristics 1, 2, 3, and 5, instead of characteristics 1, 2, 3, and 4 to show that a logical system is incomplete. This substitute characteristic turns out to be easier, though tricky, to prove for arithmetic.

Now the going gets tough. Look again at Figure 34 for help. From all possible cattle drives, choose one at random, which we will identify as the "Test Cattle Drive." Since we are not considering any specific drive, any conclusion we draw from our analysis must apply to any cattle drive at all.

We are curious as to whether or not someone will claim to be a member of this arbitrary drive under the conditions of characteristic 5. If this turns out to be true, we are safe to conclude that someone will claim to be a member of any possible cattle drive we may choose.

We learned earlier that a cowboy will only claim to be a member of a cattle drive if he is indeed a member of it and an outlaw will only claim to be a member of a cattle drive if he is not a member of it. In the present case, this means that we want to guarantee that no matter which drive we chose, there would be at least one cowboy who participated in it or at least one outlaw who did not participate in it.

The fundamental difficulty is that we cannot tell the cowboys from the outlaws unless we catch them telling the truth or lying. Fortunately we have one statement to test the veracity of a questionable individual: *A sidekick will always say that his companion is a trail boss.* Our basic strategy is to set up a situation for which you and I know its validity, and then compare the truth we know against what the individual says.

Once again, our left finger goes back on the test drive; our right one, back on the cattle drive named after the mysterious Mr. X. Now there are exactly two mutually exclusive possibilities: Either Mr. X belongs to the drive named after him or he does not. Stated differently: Either Mr. X is one of the individuals in the drive your right finger indicates or he is not. We will examine both cases.

CASE I: Mr. X is a trail boss because he belongs to the cattle drive named after himself.

That is, since Mr. X (whoever he is) is a trail boss, one of the individuals indicated by your right finger is Mr. X. Perhaps he is the cowboy #1. In any event, he has at least one sidekick back in the test cattle drive indicated by your left finger. What will this sidekick say about Mr. X? That he is a trail boss, of course, which happens to be true. The sidekick is telling the truth. The sidekick is a cowboy. The sidekick is a cowboy within the test cattle drive. The sidekick will truthfully say he is a member of the test drive.

CASE II: Mr. X is not a trail boss because he does not belong to the cattle drive named after himself.

That is, since Mr. X (whoever he is) is not a trail boss, slip your right hand down to the cowboy #4. In any event, he has at least one sidekick who is not a member of the test drive, perhaps cowboy #3. What will that sidekick say about Mr. X? That he is a trail boss, of course, because that is what a loyal sidekick always does. This happens to be a lie. The sidekick is not telling the truth. The sidekick is an outlaw. The sidekick is an outlaw outside (or in the complement to) the test cattle drive. The sidekick will untruthfully say he is a member of the test drive.

The result is clear. Whether or not Mr. X is a member of the mystery cattle drive or not, there will always exist a Texan (in our case either the one labeled #1 or #4) who will claim to be a member of the test cattle drive. But this conclusion is true for all cattle drives, not just the test case we have considered, because we started our analysis with an arbitrary drive. It follows that characteristic 6 implies characteristic 4 and our job is done. The Ghost Riders are incomplete in that there must be one cowboy who is not wearing his hat.

The system with a combination of characteristics 1, 2, 3, and 6 is really a disguised form of Gödel's famous incompleteness theorem. Fortunately for our program of trying to compel belief that arithmetic is best treated as a heuristic, they are characteristics of arithmetic if you make the correct mapping from the Ghost Riders in the Sky.

Fifth Approximation

For those who remain, the next step in studying this proof is to read the little buff-colored book, *Gödel's Proof*, by Nagel and Newman that precipitated my original crisis. It was originally written for a popular science magazine and intended for the general public. Although it is difficult, it is accessible. A very old article, "Goedelian Sentences: A Non-Numerical Approach" from *MIND* formed the basis of the consideration of language as a model for inconsistency. If even more study is desired, an excellent book by R. Carnap, *Logical Syntax of Language*, and the oddly titled book by Smullyan, *What Is the Name of This Book?*, are worth consideration. The latter inspired our tale of Lucky Luke. Of course, for those who have a lifetime to spare and are not faint-hearted, Gödel's original article, "Über formal unentscheidbare Sätze der Principia Mathematica und verwandter Systemes I" (*Monatshefte für Mathematik und Physik*, Vol. 38, 1931, pp. 173–198) is always available.

Mathematics as Mathematics

Gödel's Proof is more profoundly unsettling than has been implied so far for two reasons. First, mathematics is actually a house of cards in which the consistency of algebra, geometry, and the rest of mathematics ultimately depends on the consistency of arithmetic through a series of relative proofs. As the consistency of arithmetic goes, so goes the consistency of the edifice, mathematics. One recent author said that it is heart-rending to describe the current state of mathematical rigor. Since all of mathematics depends on arithmetic in this critical way, I also feel compelled to write mathematics as the

HEURISTIC: Mathematics

from now on. This whole matter is disturbing because mathematics has often been considered as the final redoubt by the philosopher when it comes to obtaining certain knowledge.

Deduction as Deduction

Unfortunately, the matter does not rest here for a second reason for concern exists. Gödel's Proof does not just apply within the domain of mathematics. In fact, it was not directed specifically at either, but at all (with a few minor, unimportant exceptions) axiomatic systems. This doubt about the consistency of axiomatic systems translates into doubt about the Aristotelian goal of a perfect deduction from first principles. It is a body blow to the entire scientific enterprise and the whole body of rational knowledge.

Gödel's Proof is not the only problem with deduction as a way to know about our world. Classically, a deductive argument is certified as valid if it begins with an appropriate set of premises and follows a specified procedure. As is well known, this form of argument does not claim to guarantee the truth of the conclusion, but only that the conclusion follows logically from the initial premises. The syllogism

> All men are stupid;
> Socrates is a man;
> Therefore, Socrates is stupid.

is usually considered to be a valid argument, but because of doubt about the major premise, the conclusion is not necessarily true. Very soon we will examine the issues of *logic* and *truth* and have additional reasons besides Gödel's Proof to have concern about this entire enterprise.

Based on Gödel's Proof, I feel I must write the

HEURISTIC: Deduction.

Echoing what we earlier said about language and induction, from this point on, any argument supported in part by arithmetic, mathematics, or deduction must be considered as a heuristic argument.

Gödel's Proof, and the worldview that it dictates, amazes most people when encountered for the first time. For over thirty years, my engineering and liberal arts students have expressed surprise and concern that such an important factor for determining how each person thinks about his world was not a part of the curriculum. Perhaps you feel the same way. As we shall soon see other equally amazing, little known, and unbelievable concepts remain to be examined that likewise should be taken into account as you try to understand your world.

We live in privileged times. We are extremely fortunate to live in an age when Gödel's Proof holds sway. We can use Gödel's attack on the most sacred of notions, *mathematics*, to achieve an easy, early victory in showing that everything is heuristic. Our task has been made psychologically easier because arithmetic is often thought of as the ultimate standard for how knowledge ought to be. We receive encouragement because whatever heuristics were used in the past to make the certainty of mathematics so compelling turn out to be bad ones. It would have been harder for us to get a hearing had they never led us astray.

Our claim, however, must not rest on the vagaries of time. Prior to 1931 we would have still wanted to assert that "All is heuristic," but doing so would have been more trying for both of us. All we require is that errors are possible, that all concepts are potentially fallible, not that errors are known to be currently present. We are fortunate to live at a time when our kōan cannot be met with the counterattack, "But what about arithmetic?"

In the strength of our present position, however, is borne a subtle weakness that we have met before. Perhaps, just perhaps, Gödel's Proof is one more camel's nose that we are psychologically elevating to an absolute. We may be making the same error in the case of arithmetic that we have just criticized others for making. Just as we saw earlier with the *diallelos*, an argument can become too convincing, can be thought to be an absolute, and can effectively defeat our basic premise. To be on the safe side, let's voluntarily weaken our own argument and write Gödel's Proof.

Now that we have laid down the first two warp and the first three woof threads, it is time to make a few comments to assure that our weaving is straight. If we do not, we risk being disappointed in the final hinklejawed fabric. My program to compel belief that *All is heuristic* is now obvious. I intend to examine a large number of important concepts and suggest that they are best considered as heuristics. I do not insist that they are not important, that they are not useful, that they are not the best we can do to understand our world. I only insist that though plausible, helpful, and based on experience, that they might quite possibly be fallible. How many important concepts should I consider? That depends on each individual. For some, Gödel's death blow to mathematics will suffice to compel belief in this notion. I know it did for me. For others, only an exhaustive consideration of every possible idea, concept, and creation of the human mind would be acceptable. For most, I can only hope to have enough examples to befuddle and let induction work its way.

We have a magic rug to weave; let us take up our shuttle, once again, and go back to work.

Certain as Certain

When I began this section I noted that experts could agree that a concept was in doubt or they could agree that it was certain. Arithmetic served as an example of the first case. Now we must turn to the second.

The psychological force of finding one concept that is thought *at the present time* to be absolutely certain (much as was thought of arithmetic prior to 1931) was made clear to me one day in the classroom when a frustrated student—trying to stop my monotonous repetition of the mantra *All is heuristic*—blurted out, "Challenge them to name one thing that is absolutely certain and not a heuristic. That'll shut them up."

Yes, we do live in privileged times. All efforts to find one concept that all experts can agree is certain and beyond doubt have so far failed. Because of the existence of a group of philosophers called skeptics, one is unlikely to emerge in the near future.

Let us look at some notable failures to find the certain and then make the acquaintance of these peculiar philosophers. We can then evoke the

HEURISTIC: If we are unable to find a single example of a concept, we should suspect that it may not exist.

to force us to write certain as the

HEURISTIC: Certain.

Past failures to find at least one thing that is beyond doubt have not been for wont of trying and have not dissuaded others from seeking the certain. For example, the classical view that geometry and arithmetic are beyond doubt and should serve as models for all knowledge turns out, as we have just seen, to have at least a hint of doubt because of Gödel. Likewise, as we will see later, Descartes' notion of "clear and distinct" ideas turns out to be neither clear nor distinct and has not won universal assent. The champion in the effort to produce the certain might well have been Descartes' good friend, Mersenne, who held that in every field of human interest some things were simply known. Among a wide variety of examples, he offers the assertion "the whole is greater than the part." Most people, and you may be among them, would agree with Mersenne that this is obviously, simply, indubitably so. Unfortunately, Mersenne was wrong. The whole is not always larger than its parts—at least not when infinite collections are concerned. To show this, write the natural numbers, 1, 2, 3, . . . , in one row with the even numbers, 2, 4, 6, . . . , in a second row below it so that each number in the first row is directly above the number in the second row that is its double, as follows:

1	2	3	4	5	6	7	8	9 . . .
↑↓	↑↓	↑↓	↑↓	↑↓	↑↓	↑↓	↑↓	↑↓
2	4	6	8	10	12	14	16	18 . . .

Every number in row one from 1 to infinity has a mate in row two. Since the bottom row results from removing all of the odd numbers from the top one, the part (the even numbers) is the same size, or equal to, the whole (all natural numbers). Infinite collections do not behave as a sota based on finite collections expects. This line of reasoning, made more precise and acceptable by modern analysis, is now a standard result in non-Cantorial set theory. To dismiss this example as an oversight on the part of Mersenne misses the fundamental point: A notion of *certain* based on a personal sota is based on a very unsubstantial foundation. For our purpose, it is sufficient to note that time has shown that Mersenne's example of a concept that was "simply true" is now considered to be far from simple and far from true. *Doubt, using time as a weapon, has always retaken any beach-head once thought secured by the certain.*

Position as Position

The comments of two celebrated philosophers, G. E. Moore and L. Wittgenstein, will serve as more sophisticated examples of the futile search for the certain. In a series of essays Moore claims simply to know a number of propositions. His view is that any argument proposed to refute them is more suspect than the propositions themselves. The most famous of his claims is, "Here is one hand, and here is another," uttered as he raises his hands. We will not consider this example now with the promise to examine it more closely later when we deal with perception. Moore's second self-evident proposition, "I have never been far from the earth's surface" will occupy us at this time.

In the book *On Certainty* Ludwig Wittgenstein considers Moore's comment and concludes:

> Isn't it true, or false, that someone has been on the moon? If we are thinking within our system, then it is certain that no one has ever been on the moon. Not merely is nothing of the sort ever seriously reported to us by reasonable people, but our whole system of physics forbids us to believe it.

Our immediate concern is with the phrases *seriously reported* and *physics forbids*.

Why, in this passage, did Wittgenstein imply that *our* system of physics forbids him from believing that he had been on the moon? *His* system of physics (that is, his sota) seems to have forbidden it, but the sota of the modern physicist does not. Like Mersenne's, Wittgenstein's *certain* based upon his personal sota is far from an absolute one.

Wittgenstein continues, "Why is it not possible for me to doubt that I have never been on the moon? And how could I try to doubt it?" The answer to the first question is profound and an essential part of this discussion. Remember it well; we will discuss the compelling nature of a sota later. The answer to the second question is simpler. Wittgenstein could begin to doubt by recognizing his sota for what it is, a very personal collection of heuristics, and by studying the sota of modern physics.

This we will do now with the caution that my rejection of Wittgenstein's feeling of certainty in this quotation is not based upon the trivial criticism that in 1951 he could not foresee the moon landing in 1969. His philosophical analysis is far more subtle than a superficial improvement in technology can refute. If he were writing today, he would simply choose another location, say, by replacing the moon with Alpha Centauri. My rejection is based on the curious way in which his feeling of absolute certainty is based on his personal sota. To make you feel the force of my rejection, I ask you, "Do you feel that it is certain that you have never been on Alpha Centauri?" If you do feel that it is certain, "Why?" and "How could you try to doubt it?"

Quantum mechanics is one of the most widely accepted theories in modern physics. The fundamental concept in this theory is the solution to a complex mathematical equation known as the Schrödinger wave equation. The exact nature of the wave equation need not concern us now, but to satisfy my colleague's earlier demand that every book contain at least one mathematical equation, it is given here:

$$\frac{\delta\psi^2}{\delta x^2} + \frac{\delta\psi^2}{\delta y^2} + \frac{\delta\psi^2}{\delta z^2} + \frac{8\pi^2 m}{h^2}\ [W - V]\Psi = 0$$

The Greek letter psi, or Ψ, represents the fundamental quantity of quantum mechanics. It contains all of the relevant information about a specific system or object. Given Ψ, a mathematical operation (related to squaring its amplitude) creates a second mathematical quantity called a relative probability density function, represented mathematically as

$$\int \psi\psi^* d\tau.$$

The relative probability density function represents the probability of finding the object whose wave function is ψ at various locations in space. Many physicists believe that Wittgenstein can be represented by a ψ and that both ψ and $\int \psi\psi^* d\tau$ for Wittgenstein are defined over all space. In particular, these two quantitites have a small, but calculable, value on the moon and on Alpha Centauri. The results of modern physics insist that Wittgenstein's implied question, "Am I or have I ever been on the moon?" cannot be answered by true or false. The best that can be said is that Wittgenstein is "mostly here," but also "partially on the moon." We do not expect to see humans appear spontaneously on the moon or Alpha Centauri because of the extremely small value of the relative probability density function of a human in these locations, not because it would violate the laws of physics.

This view of Nature is not just the ravings of theoretical physicists who have lost all touch with reality. In a form of radioactivity called alpha decay and in the tunnel diode used by the electrical engineer, experimental evidence conclusively demonstrates the validity of the physicist's theory. Once again, if you do not have a lifetime to spare for the study of quantum mechanics to verify these results for yourself, you will have to believe one who has had experience, that is, *experto credite.*

The relative probability density function of quantum mechanics is not the only result of this grand theory that casts doubt on Wittgenstein's claim to have never been on the moon. Consider the Heisenberg Uncertainty Principle, which casts doubt on our ever knowing exactly where we are.

The Heisenberg Uncertainty Principle is accepted by a near consensus of modern physicists and states that a fundamental limitation exists in the ability to measure certain physical variables. Energy, time, momentum, and position are examples of these quantities. To be specific, consider the pair of variables, position and momentum. The Heisenberg Uncertainty Principle requires that the more accurately the momentum of, say, an electron is known, the less accurately its position can be specified and vice versa. In other words, we can never know with certainty the position *and* the momentum of anything. But note this well: The imprecision we observe in the electron's position is not due to a failing of the human's measuring technology, but is a fundamental characteristic of the world. Indeterminacy is ingrained in Nature. Direct quotations of Robert Oppenheimer and George Gamow support this paradoxical point:

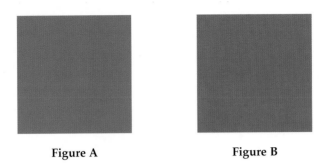

Figure A   Figure B

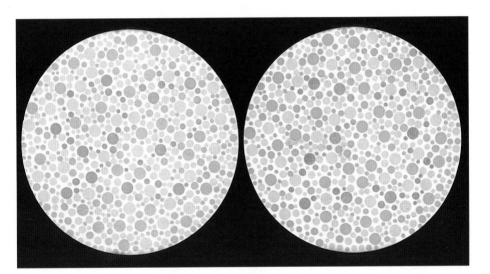

Figure C A Test of Color Blindness

Figure D

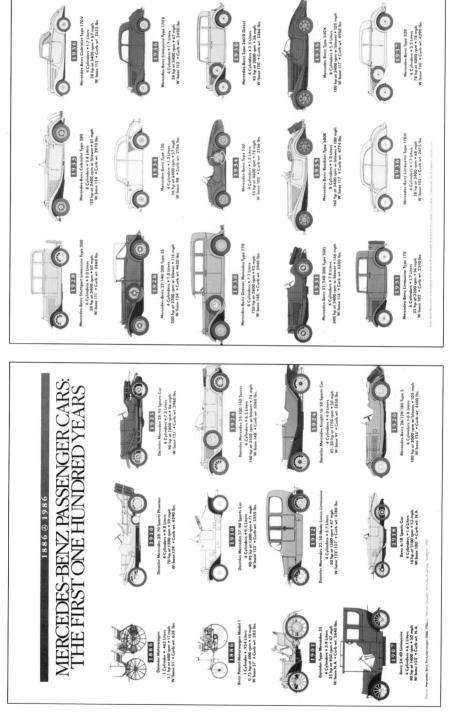

MERCEDES-BENZ PASSENGER CARS:
THE FIRST ONE HUNDRED YEARS

1886 ⊕ 1986

1886
Daimler Motorwagen
1 Cylinder • .462 Liters
1.1 hp at 600 rpm • 11 mph
W.base 51 " • Curb wt. 628 lbs.

1886
Benz Patent-Motorwagen Model 1
1 Cylinder • .954 Liters
0.75 hp at 400 rpm • 10 mph
W.base 57 " • Curb wt. 583 lbs.

1901
Daimler Type: Mercedes 35
4 Cylinders • 5.9 Liters
35 hp at 950 rpm • 47 mph
W.base N.A. • Curb wt. 2640 lbs.

1907
Mercedes 24/40 Limousine
4 Cylinders • 6.1 Liters
40 hp at 1300 rpm • 50 mph
W.base 123 " • Curb wt. N.A.

1910
Daimler Mercedes 38/70 Sports Phaeton
4 Cylinders • 9.8 Liters
70 hp at 1200 rpm • 59 mph
W.base 139 " • Curb wt. 4290 lbs.

1910
Daimler Mercedes 37/90 Sports Car
4 Cylinders • 9.5 Liters
90/95 hp at 1300 rpm • 71 mph
W.base 133 " • Curb wt. 3355 lbs.

1912
Daimler Mercedes 22/50 Mode-Luxus-Limousine
4 Cylinders • 5.7 Liters
50 hp at 1300 rpm • 47 mph
W.base 123 127" • Curb wt. 4180 lbs.

1918
Benz 6/18 Sports Car
4 Cylinders • 1.6 Liters
18 hp at 2100 rpm • 53 mph
W.base 100 " • Curb wt. N.A.

1921
Daimler Mercedes 28/95 Sports Car
6 Cylinders • 7.3 Liters
90 hp at 1800 rpm • 86 mph
W.base 121 " • Curb wt. 3960 lbs.

1924
Daimler Mercedes 24/100/140 Tourer
6 Cylinders • 6.2 Liters
140 hp at 3100 rpm • 74 mph
W.base 148 " • Curb wt. 5060 lbs.

1924
Daimler Mercedes Knight 16/50 Sports Car
4 Cylinders • 7.0 Liters
45 – 50 hp at 1750 rpm • 50 mph
W.base 91 " • Curb wt. 3850 lbs.

1926
Mercedes-Benz 26/120/180 Type S
6 Cylinders • 6.8 Liters
180 hp at 3000 rpm • 105 mph
W.base 134 " • Curb wt. 4400 lbs.

1928
Mercedes-Benz Stuttgart Limousine Type 200
6 Cylinders • 2.0 Liters
38 hp at 3400 rpm • 50 mph
W.base 111 " • Curb wt. 2860 lbs.

1928
Mercedes-Benz 27/140/200 Type 55
6 Cylinders • 7.0 Liters
200 hp at 3200 rpm • 115 mph
W.base 134 " • Curb wt. 4620 lbs.

1930
Mercedes-Benz Grosser-Mercedes 770
8 Cylinders • 7.7 Liters
150 hp at 2800 rpm • 93 mph
W.base 148 " • Curb wt. 5940 lbs.

1931
Mercedes-Benz 27/240/300 Type SSKL
6 Cylinders • 7.0 Liters
300 hp at 3400 rpm w/blower • 146 mph
W.base 116 " • Curb wt. 3300 lbs.

1931
Mercedes-Benz Limousine Type 170
6 Cylinders • 1.7 Liters
32 hp at 3200 rpm • 56 mph
W.base 102 " • Curb wt. 2310 lbs.

1933
Mercedes-Benz Cabriolet Type 380
8 Cylinders • 3.8 Liters
120 hp at 3400 rpm w/blower • 81 mph
W.base 124 " • Curb wt. 2970 lbs.

1934
Mercedes-Benz Type 130
4 Cylinders • 1.3 Liters
26 hp at 3400 rpm • 57 mph
W.base 95 " • Curb wt. 2156 lbs.

1934
Mercedes-Benz Type 150
4 Cylinders • 1.5 Liters
55 hp at 4600 rpm • 78 mph
W.base 102 " • Curb wt. 2156 lbs.

1935
Mercedes-Benz Roadster Type 500K
8 Cylinders • 5.4 Liters
160 hp at 3400 rpm w/blower • 100 mph
W.base 117 " • Curb wt. 4774 lbs.

1936
Mercedes-Benz Limousine Type 170H
4 Cylinders • 1.7 Liters
38 hp at 3400 rpm • 78 mph
W.base 107 " • Curb wt. 2475 lbs.

1936
Mercedes-Benz Cabriolet Type 170 V
4 Cylinders • 1.7 Liters
38 hp at 3400 rpm • 67 mph
W.base 112 " • Curb wt. 2552 lbs.

1936
Mercedes-Benz Limousine Type 170 V
4 Cylinders • 1.7 Liters
38 hp at 3400 rpm • 67 mph
W.base 112 " • Curb wt. 2420 lbs.

1936
Mercedes-Benz Type 260 D Diesel
4 Cylinders • 2.5 Liters
45 hp at 3000 rpm • 56 mph
W.base 120 " • Curb wt. 3386 lbs.

1936
Mercedes-Benz Type 540K
8 Cylinders • 5.4 Liters
180 hp at 3400 rpm w/blower • 105 mph
W.base 117 " • Curb wt. 4950 lbs.

1937
Mercedes-Benz Type 320
6 Cylinders • 3.2 Liters
78 hp at 4000 rpm • 78 mph
W.base 130 " • Curb wt. 4290 lbs.

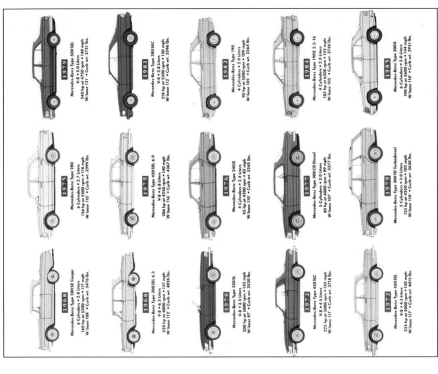

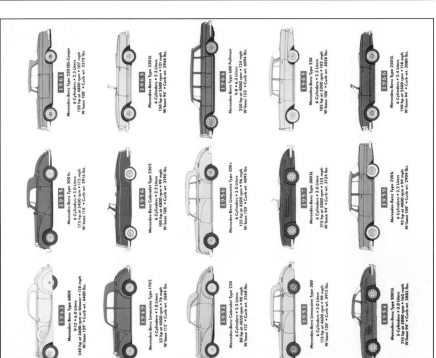

Evolution of the Mercedes Automobile

Lucky Luke

The Ghost Riders in the Sky

If we ask, for instance, whether the position of the electron remains the same, we must say "no"; if we ask whether the electron's position changes with time, we must say "no"; if we ask whether the electron is at rest, we must say "no"; if we ask whether it is in motion, we must say "no".

You cannot actually indicate the position of the electron exactly, the best you can say is that the electron is "mostly here" and "partially somewhere else". You may think that this is unusual but the commentator continues, "On the contrary it is absolutely usual, in the sense that it is always happening to any material body."

Since Oppenheimer and Gamow are very well-known and respected scientists, I feel justified in bringing the following

HEURISTIC: If direct quotation from creditable sources suggests caution in accepting the validity of a concept, exercise caution yourself

to my defense. The Heisenberg Uncertainty Principle and $\int \psi_i \psi_i^* d\tau$ serve as the vehicle to transport Wittgenstein to the moon.

Given these two results, I seem to have no recourse but to write:

HEURISTIC: Position.

To be on the safe side, we should probably write momentum, energy, and, our old friend, time to indicate the status of the other variables that come under the sway of the Uncertainty Principle. When in doubt, labeling a concept a heuristic is always a safe heuristic.

Once again, I do insist, "We live in privileged times." Before the advent of quantum mechanics, we could not have even conceived of the notion that we could have been on the moon. Witness Wittgenstein's feelings of disbelief. But then again, remembering our concern over the *diallelos* and Gödel's Proof, shouldn't I have just now written *quantum mechanics*? I cannot afford to make my argument too convincing.

Descartes, Mersenne, Moore, and Wittgenstein are but a few on a long list of philosophers who have sought one object that a consensus of experts would agree is certain—and failed. Finding expert agreement on a future success to replace these earlier failures does not look promising. Among philosophers there exists a group of professional doubters, the skeptics, who have won every battle they have fought. Of course, I mean by a skeptic what everyone else does, a person who doubts everything, one who questions the possibility of real knowledge of any kind. I mean that person who will argue that snow is black if you argue that it is white or the converse since it is all one to him on which side of a debate he finds himself. At one time or another, the skeptic has doubted:

1. All knowledge claims except the claim that he doubts all knowledge claims (academic skepticism)
2. All knowledge claims including the claim that he doubts all knowledge claims (Pyrrhonism)
3. The data of sense
4. The existence of all but the self (solipsism)

5. All but religious faith
6. All facts that go beyond experience
7. That immediate experience is possible
8. Reality or the real world
9. That anything can be self-evident
10. That we can tell which of our judgments are true
11. All that go beyond appearance

Obviously this list is not comprehensive, but it does show the wide variety of often contradictory skeptical thought.

Skepticism has not been a minor, unimportant philosophical position. Its adherents have been among the most famous, thoughtful, and influential scholars who have studied the theory of knowledge. This is so if we can believe Cicero, who told us that "almost all the ancients have said that nothing can be understood, nothing perceived, nothing known; that our senses are narrow, our minds weak, the course of our life short." This is so if we can believe Gorgias, who said, "Nothing exists; if it did, we could not know it; if we could know it, we could not communicate our knowledge to others"; or if we can believe Xenophane, who said, "All human knowledge is guesswork"; or if we can believe Montaigne, who reminded us, "The historical, the dialectical, the physiological, the psychological arguments do indeed show that the human mind is unable to attain absolute certainty in any field."

Now I lay it down as a fact that I am also a skeptic. Not only am I a radical skeptic, but the heuristic position we are now discussing will soon be found to be even more radical than Pyrrhonism, one of the most radical brands of skepticism known. Will I doubt everything? Will I doubt that I exist? I can go as far as that and even farther. Nature does not prevent my raving to this extent. I gladly join a large family of skeptics, although not through the front door as a philosopher, scientist, or mathematician, but through the delivery entrance as an engineer. Among my philosophical ancestors are Pyrrho, Timon, Arcesilaus, and Carneades, who were put on the skeptic's family tree by the genealogist Sextus Empiricus. In every family there must be an eccentric uncle. Because he was so peculiar, the skeptic Agrippa will play this role. My intellectual fathers are Descartes, Pascal, Bayle, Hume, and, although he tried to deny the relationship, Berkeley. For brothers I have Santayana, Camus, and perhaps Popper. But even this extensive list does not do justice to the size of my family. I am convinced that most philosophers today are at least covertly skeptical about the possibility of attaining certain knowledge. Because of their number and the effectiveness of their arguments, the skeptics have been a major force in epistemology throughout history. Their success continues even to the present day. A person, such as I, who believes that everything is a heuristic, is proud to claim to be one of their blood relatives. The only fault I find with my kin is that they are far too timid.

Later we will examine the difference between the skeptical and heuristic positions to see how truly radical we have become. For now, we must recognize the existence of well-known scholars who are prepared to doubt anything put forth as certain. The power of the skeptic is such that it seems unlikely that past failure to achieve expert agreement on the certainty of any concept will be replaced by a future success.

Since it is hard to take seriously the existence of any concept for which no exemplar exists and for which one is unlikely to exist in the future, I feel obligated to treat certainty as a heuristic.

The last warp heuristic we considered focused on the agreement of experts. We have also considered the woof heuristics *arithmetic, mathematics, deduction, certain,* and *position* once taken to be certain whose status has become less certain. Now attention shifts to disagreement, to the

HEURISTIC: If qualified experts in the *same* field hold widely differing opinions on a matter, then we should suspend judgment.

The concern expressed in this heuristic is at least as old as Agrippa, who specifically listed the conflicting opinions of experts as one of his five skeptical arguments for rejecting the arguments of the dogmatist. The woof heuristics *logic* and *truth* will illustrate the force of this heuristic.

Logic as Logic

We like to think that even if we cannot define what we mean by "logical," at least the expert can. We are badly mistaken. Depending on different philosophical and cultural leanings as to the nature of mind, reality, and knowledge, different logical systems develop.

Lists are dull. If I only wanted to demonstrate that on occasion experts disagree about the important matter of logic, perhaps a list of differing opinions could be avoided. But my purpose is to emphasize the wide range of crucial areas in which there is expert disagreement, and to do so, not treatises on several selected areas but a list of a large number of diverse areas is more appropriate. You are free to skim the following partial list of different logic systems:

1. *Laws of Thought* They consist of (1) the law of Identity (Whatever is, is); (2) the Law of the Excluded Middle (Everything must either be or not be); and (3) the Law of Contradiction (Nothing can be and not be). It appears that no viable logic system can be constructed in which these laws are the only axioms.

2. *Aristotelian Logic* This system gives a satisfactory account of a judgment and reasoning, but fails to give a good account of mathematics.

3. *Symbolic Logic* It may succeed (but note Gödel) in formulating the structure of modern mathematics, but not certain metaphysical systems.

4. *Formal Theory of Probability* This is an infinite valued logic with truth-values ranging from zero to one that does not have the law of the excluded middle.

5. *Multivalue Logic* This logic system is Łukasiewicz's and is based on three logical values: the true proposition, the false proposition, and the constant "possible" proposition. With the discovery of n-value logic, the scientist was forced to recognize that he might misrepresent reality by arbitrarily using the classical bivalent system. It lacks the law of the excluded middle and the law of contradiction.

6. *Hegelian Logic* This system is Hegel's and is based on the dialectic: thesis, antithesis, and synthesis. This historical doctrine not only rejects the

law of contradiction, but declares that contradiction is inherent in the nature of reality.

7. *Quantum Logic* This system belongs to von Neumann and Birkhoff, who demonstrated (using a familiar phenomenon to physicists, the polarization of light) that it is impossible to describe experience with classical logic. It rejects a fundamental consideration in classical logic known as the distributive law.

8. *Other Logical Traditions* When we think of logic and by implication assign it a privileged status in the description of reality, we naturally use our Western sota as a basis. Increased familiarity with other traditions (notably the Indian and Chinese) reveals other logic systems uninfluenced by Western logic. For example, we have already discussed the peculiarity of the Chinese sentence as a string of uninflected ideographs that is inconsistent with our traditional concepts of predication. As a second example, Zen Buddhism is a philosophy of utter skepticism that finds contradiction intrinsic to nature and being.

Even this abridged list shows the wide diversity of expert opinion concerning the underlying logic of reality. How would these same experts have you and me choose among this offering—surely not logically? At best, whatever the selection we make is, it is the

HEURISTIC: Logic.

The heuristic nature of logic is not only supported by the wide variety of dissimilar systems that have been proposed, but another heuristic that we have used to good effect in the last part will also come to our aid. I refer, of course, to the

HEURISTIC: If a concept produces paradoxes, unexplained complexities, or departures from expected results, better consider it a heuristic.

We used this heuristic effectively when we considered the general public's concept of time. In that analysis we encountered numerous paradoxes in our usual way of looking at time, such as the twins who age differently, the clock that stops when we go at the speed of light, and the one that runs more slowly on the sun. We also used the same heuristic with induction when we considered the grue emeralds and the nonblack nonravens. Let us return to this heuristic for one more example to raise concern about the absolute nature of logic. I refer to an example by Church, who proved that no general decision procedure can exist that will tell whether or not an arbitrary well-formed formula is a theorem of elementary logic. As an example of the kind of problem in logic that seems answerable but falls under Church's proof, we have the following: Given that all horses are animals, may we infer that all heads of horses are heads of animals? Church says "no"; yet we all know it to be true. Anyone must surely find that a strange, perplexing situation.

Whether by examining a partial list of expert opinion or by considering the paradoxes of logic, it becomes clear that the principles of logic, any logic, are not *a priori* truths, but intuitive cultural constructs, heuristics I would say, for the fulfillment of human purposes. I, for one, intend to continue to admonish my two sons, Kent and Doug, to be logical, and to be glad my wife, Deanne, thinks logically.

Earlier I insisted that we should consider any argument based on induction, language, arithmetic, mathematics, or deduction as a heuristic argument. Now I am afraid I must add to this list any argument based on logic as a heuristic argument also.

Truth as Truth

"What is truth?"

This question has been asked for so many centuries, by so many people, and in so many ways that it has become, along with "What is life?" a genuine philosophical cliché for the futility of all knowledge. Our next warp

HEURISTIC: If the literature treating a concept is enormous, we should be suspicious that the matter is far from settled and that we are dealing with a heuristic

will now serve us well.

If lists are dull, how much duller is a list of lists? The research into the nature of truth is so extensive that such a list of lists is necessary to closely juxtaposition the many divergent theories as to what truth is and to bring doubt to the matter. Now, more than before, feel free to skim. The following four, occasionally overlapping, lists will be given to show the variety of expert opinion on the closely related topics:

1. Truth
2. True
3. Theories of truth
4. Criterion of truth

Whether we comb through what the experts have said about *truth*, that it is

1. An object of animal faith, not pure contemplation
2. A metalinguistic property of sentences
3. Subjectivity
4. Limited to the ongoing process of human beings as thinking subjects
5. Self-evident (impossible to prove, but also impossible to disbelieve)
6. Ultimately an act of will (existing only as it is manifested in authentic belief)
7. What answers a vital need by removing the feeling of insecurity and perplexity (it is always relative to particular life situations and historical periods)
8. An attribute, not so much of discourse, as reality
9. A matter of correspondence between thought and object, but object is not something transcendent; it is simply an immediate cognition—truth is a relationship between two levels of cognition

10. Impossible as something absolute, timeless, and pre-existent to our choices

11. Not a property of propositions but of complexes of questions and answers, complexes that rest on "absolute presuppositions" that are neither true nor false

12. Inferior to the notion of practical justice and should be suspended or even denied if practical justice and theoretical truth conflict

13. Abolished as objective, immutable, or eternal

14. A concept that comes in two classes: relations of ideas and matters of fact

15. Not an absolute notion but a relative one

or whether we consider what has been said about its close relative, *true*, that it

1. Applies to hypotheses that work

2. Is a property of our beliefs, not things

3. Cannot be defined but must be immediately recognized

4. Is not a relationship between a proposition and something outside of it

5. Is only the expedient in our way of thinking

6. Is a straightforward adjective applying to straightforward objects—sentences

7. Is to say of what is that it is, and of what is not that it is not

8. Is used for purposes of emphasis or style or to indicate the position of a statement in an argument

9. Is to perform the act of agreeing with, accepting, or endorsing a statement

10. Is meaningless and should be removed from the English vocabulary

11. Has no independent meaning but can be used only as a component of the two expressions "it is true that" and "it is not true that"—with each of these two expressions treated as single words with no parts

or whether we discuss the various *theories of truth*, such as the

1. Correspondence theory—truth consists in some form of correspondence between belief and fact

2. Coherence theory—to say that a statement is true is to say that it coheres with a system of other statements, that it is a member of a system whose elements are related to each other by ties of logical implication

3. Pragmatic theory—truth is preeminently to be tested by the practical consequences of belief

4. Existence theory—true belief is directed toward what is; false belief is directed toward what is not

5. Performance theory—to say that a statement is true is not to make a statement but to perform the act of agreeing with it

6. Nihilistic theory—the word *true* has no independent meaning but can be used as a component of the two meaningful expressions, "It is true that" and "It is not true that"

7. "No truth" theory—to say a proposition is true means no more than to assert the proposition itself

8. "Long run" theory—truth is the opinion that an indefinite community of scientific investigators will ultimately agree upon after continued experimental inquiry

9. Biological theory—truths of inferior value are eliminated, while truths of superior value survive; the belief that proves to have more survival value shows itself to be the most useful and so the most true

or, finally, whether we recognize the *criterion of truth* (that sign by which we know truth even if we do not know what truth is) as

1. Not universal—each individual is "the measure of all things: of what is, that it is; of what is not, that it is not"

2. The *consensus gentium*—the general agreement or consensus of the people

3. Natural law—a set of timeless truths that all men could read for they are "engraved on their hearts in letters more durable than bronze"

4. The "successful working" of an idea—the idea that works is true

5. Non-falsification (either in the sense of the Western tradition or the Indian one)—the latter exemplified by Sankara, who felt that there were two levels of truth: Ordinary, commonsense knowledge was considered as veridical and recognized as truth up to the moment when it was specifically falsified by the higher experiences of Brahman

6. The "clear and distinct"

7. Nonexistent

we are no closer to the meaning of truth. Strand upon strand, list upon list, heuristic upon heuristic, who will unravel this tangle? Who will free me from the tyranny of the uncertainty of truth?

Only I can break my shackles by writing the

HEURISTIC: Truth

to go with the heuristic *logic*. Having done so, I repay the promise made earlier and cast final doubt on the absolute nature of *deduction*.

Progress as Progress

Following the lead of Agrippa, lists of conflicting expert opinion have been offered as corroborating evidence that the philosophical concepts of logic and truth should be treated as heuristics. Some may object to this procedure. Do not isolated selections of thought ignore the progress by which an idea, once accepted as correct, is later superseded? In much the same way that science assumes it can make progress and know that it does so, maybe one philosophical view of logic or truth could be identified as absolutely true or at least the best current approximation to truth. Perhaps—but the idea of progress creates its own problems both in philosophy and, as we will later see, in science.

First, progress must also be considered a heuristic for reasons already examined. Almost as much controversy, literature, and research have surrounded the concept of progress in modern times as has surrounded the concepts of logic and truth in the past. Once again we could easily follow the counsel of Agrippa and produce a sufficiently long list of conflicting claims to establish progress as a heuristic.

The second problem with progress, assuming now that we are dealing with a

HEURISTIC: Progress,

is that any two interacting heuristics, in this case *progress* and either *logic* or *truth*, would seem to require a third heuristic to tell us which of the first two can legitimately count as support for the other. Descartes' classic heuristic to deal with situations like this, (ironically billed as one of the rules for making progress in thought), is to "think in an orderly fashion, beginning with the things which are the simplest and easiest to understand, and gradually and by degrees reaching toward more complex knowledge." Which, then, is simpler: *progress* or *truth*?* Although selecting one notion as absolute truth from a list of contenders using the far more tenuous concept of progress runs contrary to Descartes' counsel, it might be acceptable as a heuristic, I suppose, for using the "less clear to enlighten the more clear" is a common enough human heuristic for tricking an opponent into agreeing with us. We will meet this oft-used heuristic again later, once in the hands of a master. But in the present case, where a definition of progress can hardly be constructed that does not depend, at least implicitly, on logic or truth, use of this heuristic is at best a rather peculiar tactic.

A third problem with using progress to pick a winner between rival philosophical systems is that progress can be a bit presumptuous, especially when the champions of the losing theories are unavailable for rebuttal. Even if a consensus of contemporary philosophers were convinced that progress made a *two-level version of truth* or *a historical doctrine of logic* untenable, I suspect that Sankara and Hegel would disagree and still support their respective visions. The ancient skeptics gave us the heuristic that it was unfair for one of the participants in a dispute to set the rules of the dispute. Their principal concern was with the debate between a rational and irrational worldview in which they insisted that the rationalist's demand for rational arguments on both sides was unfair. How much more unfair is it to use *progress* to establish a preferred notion of truth when some of the participants are dead? These problems are severe enough to force

---

*We cannot stop here to survey the enormous literature surrounding the concept *simpler* or to determine if *simpler* is simpler than *truth*, *logic*, and *progress*, although to do so would demonstrate the difficulty in securing the first stitch in a rational argument. We will return to the matter later when we examine the nature of science and Occam's razor.

me to reject progress as a criterion for choice between the items on a list of expert disagreement, although in an emergency I might accept progress as one.

Even with this lack of consensus among experts and the absence of progress as a criterion to resolve the disputes, I do not advocate discarding logic and truth from our lexicon; I advocate discarding them as absolute and reinstating them as absolute. Our frustration in knowing what they are, or even if they are, does not lessen their value. The complex, interrelated heuristics that surround logic and truth form a patch interwoven into the sotas of the Western tradition whose abrupt removal would unravel and render unusable the fabric of our world. They make our sota comfortable, serviceable, and familiar like an old, knitted sweater. By uttering the words *logic* and *truth*, we boldly announce to the world our view that in the matter at hand, we believe that they are good, applicable heuristics; but, by recognizing them as heuristics, we achieve liberation from the self-inflicted tyranny of these concepts. If it is any comfort, I, for one swear by Jove that I really hold the opinion that logic and truth are heuristics—and I seem to be able to make do.

Let's stop a minute to get our bearings. This is an appropriate moment for review, for we have worked our way down the first eight warp heuristics out of the eleven on the first list and considered the first eight woof heuristics out of the fourteen items on the second list. An entr'acte will be welcome to get a bit of air, to recall our goal, and to review the strategy we are using to achieve it.

We seek support for the claim that *All is heuristic*. The usual strategy to gain acceptance or compel belief in the Western world is to propose a rational argument: premises becoming conclusions through a series of steps that conform to accepted procedures. In the present case, this strategy deserves to fail, for in a world in which everything is a heuristic, the initial premises, this linear approach, the conclusions—even the language, words, and syntax in which they are expressed—become suspect and can no longer be allowed to go unchallenged. Mindful of this problem, other heritages, the Zen Buddhists as a case in point, seek understanding or enlightenment in other ways. The strategy we are using lies between those of the West and the East. Individual heuristics and the interactions of individual heuristics are reviewed in a random, almost haphazard order. Some of them seem arbitrary, overlapping, and vague; others represent the pillars that support Western thought. As with the Chinese language, conviction that *All is heuristic* resides in the emotional conclusions of the chance conjunctions of these heuristics.

As already explained, this hybrid strategy has an advantage. Unlike a traditional argument that falls apart if any step in the development is removed, the present one suffers very little if one of the heuristics it uses proves weak. As an example, in the list of various logic systems that was given earlier, a good case could be made that the English word *logic* is not being used in the same sense in all of the entries. Does the term *Hegelian logic* use the word *logic* in the same way as the term *quantum logic*? No harm is done to the overall argument if either item is removed. Indeed the original list included an additional entry, the *intuitionist logic* of Kronecker and Brouwer, but because of possible confusions over the technical word *intuition*, it was omitted. For each person the argument given here is as strong as the heuristics he has in his own personal sota.

Thus refreshed, let us return to our tapestry to consider several heuristics that make the shock of Gödel's Proof pale by comparison.

Causality as C*a*usality

When we first introduced the engineering method, we suggested that the engineer *causes* the best change in an uncertain situation within the available resources. At that time, I admitted that the use of the word *cause* was suspect and gave the implicit promise to reconsider it later. Now is the time to do so.

In life the concept of causality is so clear, so obvious, and so compelling. You know, of course, what I mean by causality. When the cue ball in a game of pool hits one of the object balls, it *causes* the latter to move. Our focus is the innocuous-appearing word *cause*. The amount of research, thought, and discussion trying to explain the relationship between cause and effect is truly enormous. One compilation of the greatest books of the Western world that appeared in 1952 lists over twenty-two hundred citations to the notion of causality, and it does not include research of more modern origin. With only mild exaggeration, in all of these references David Hume has had the first and last word on the subject, and curiously that word was that causality does not exist.

Before Hume, philosophers felt that there was a certain necessary or inherent connection between cause and effect. We might say the cause had the power to provoke or tease out a specific effect. The scientific conception was that a given stimulus under controlled conditions must inevitably produce standard results. Hume held differently. He felt that we have no evidence of any power exerted by the cause on the effect, that there was no necessary connection between cause and effect, and that causes and effects are merely changes that we find constantly conjoined. By necessity he insisted on logical necessity, and then asked what kind of logical necessity could hold between distinct, separate objects. Hume gave many examples of a relationship that would be labeled causal: fire burns, water suffocates humans, impulse produces motion, and snow causes cold. In each case, he insisted that we simply have two kinds of objects conjoined: flame and heat, snow and cold, and so forth.

Although Hume accurately pinpoints the problem of causality, to my mind, he is less successful in providing a resolution. He is a master at using a common enough heuristic we have already met. He uses the "less clear to enlighten the more clear." When we ask, "What makes us think that there is a power inherent in the cause to produce the effect?" Hume answers, "Repeated observation of the association of events leads us to the habit of expecting the association to continue," and "If flame or snow is presented anew to the senses, the mind is carried by custom to expect heat or cold." But how are *habit* and *custom* any better understood, any more clear, and any less heuristic than is causality? I, for one, see, feel, and use cause and effect daily in my engineering work, but I'm afraid I don't know what the heuristics *ha̱bit* and *cu̱stom* really mean.

Although we just sought to demote causality to a heuristic using the vast, indecisive literature that surrounds it, once again most of the previous warp heuristics and many of those warp heuristics to follow are equally effective in establishing causality's assignment as a heuristic. For example, disagreement among philosophers is rife. Aristotle accepted the notion of cause and divided it into material, formal, efficient, and final causes, but other philosophers consider it worthless, an anthropomorphism, or, in the case of the Hindus, nonexistent. Direct quotations of experts challenging the concept are also available. Bertrand Russell calls it a "relic of a bygone age," and the skeptic, Aenesidemus, in his etiological tropes, gives no less than eight ways of attacking dogmatic theories of causation.

Interestingly enough, the attacks on causality do not only come from people in closely allied disciplines. Unexpectedly, both the philosopher, as we have just seen, and the physicist have had a hand in the matter and both agree that causality is a shaky, perhaps even nonexistent, concept. To account for this situation, let me introduce a new warp

HEURISTIC: If experts in widely *different* disciplines agree that a concept is in question, perhaps we should do the same

because it will add strength to our growing conviction that causality is a heuristic. Quantum mechanics again leads the charge, this time using Einstein's EPR* thought experiment as the lance and Bell's Inequality† as the gladiator's ball and chain. We will examine each of these results from modern physics to support the philosopher's feeling that causality does not exist.

If you think Gödel's Proof was surprising and if you were disturbed that it was not a fundamental part of your worldview, wait until we review Bell's Inequality. It is such a breathtaking experimental attack on causality and perhaps the whole Western notion of reality that it deserves an explanation here.

When we considered Gödel's Proof, we ran into difficulty because the complex mathematics that was needed to derive it was denied us. As a result we had to make do with only an indication of its import. Now we find ourselves in an even more difficult situation because both the mathematics and the physics needed to understand the EPR thought experiment and Bell's Inequality are beyond our reach. Once again, we will proceed by successive approximations and, once again, I will depend on the generosity of the expert. Of course, you can skip to the conclusions on page 163 by whispering the now familiar heuristic *experto credite* at any time.

First Approximation

Quantum mechanics makes comments and draws conclusions about a strange menagerie of beasts, such as electrons, photons, quarks, and so forth, that have equally strange characteristics, such as spin, charm, color, and so forth. Although these words may seem familiar enough, they actually have very specialized meanings and to be on the safe side, I will fabricate a very fanciful example to help us begin to understand what is at issue. The following analogy is so imperfect that it might strain the good will of the expert, but it should help the complete neophyte understand the basic situation without his straining with new terminology, new mathematics, new concepts, and new physics. Our modest objective is to represent the basic concept of the EPR experiment, not to understand its exact formulation.

Consider Siamese twins who were joined at birth and who are now playing a simple game. Further assume that the game has two rules that must be obeyed at all times.

RULE 1: When one of the twins raises his hand, *instantaneously*, the other one must lower his; when the first lowers his hand, the second must raise his.

---

*Named after Einstein and his colleagues, Podolsky and Rosen.
†Bell's Inequality is also known as Bell's Theorem, Bell's Paradox, or Bell's Proof.

That is, at all times the hands of the twins must point in the opposite directions. To be on the safe side, we test whether or not this rule is being obeyed as often as we like and never find an exception. When we whisper to one twin to raise his hand and he does, immediately down goes the hand of the other. We might even be willing to say that one event caused the other: The two actions are local, in some sense conjoined, and perhaps over time we will come to have the habit of expecting the second when the first occurs. To make this notion more concrete, we could assign the number +1 to a hand that is raised and a −1 to a hand that is lowered. Rule 1 says that the sum of the numbers that represent the states of the two hands must always equal zero. It is simply against the first rule for both hands to be in the same direction or for the sum of the numbers assigned to the hands to be anything other than zero at any time.

Now let us separate the Siamese twins, leaving one in, say, Thailand and moving the other one halfway around the world. The second rule of the game now comes into play.

RULE 2: When separated, the twins can only communicate with each other at a finite speed.

What this rule means is that under no circumstances can a signal get from one twin to the other faster than whatever finite speed the game allows. As a concrete example, if they are to communicate by sound waves in the game, the maximum speed is perhaps seven hundred miles per hour. Again, we do repeated experiments and always find that the signal from one to the other never goes faster than the one designated in the game.

When the twins are connected all goes as expected; when they are separated we cannot see how both of these rules can be obeyed simultaneously. How could an event in one place cause an event in another distant place? How can an action at one place *instantaneously* cause an action at a distant one faster than communication will allow? We do an experiment. When they are separated, we whisper to one twin to raise his hand and—to our utter amazement—we find that the other one instantaneously lowers his hand as required to keep the sum zero faster than the message to do so could possibly reach him. This is certainly not causality as we have come to expect it. Basically Bell's Inequality allows experimental proof that in our world a situation similar to our twin's game is true. John Gribbin in his book, *Schrödinger's Kittens*, quotes Nobel prize-winning physicist, Brian Josephson, as saying that the experimental proof that Bell's Inequality is violated in the real world is the most important development in physics in recent times.

If you decide to move to the conclusions at this point, please promise never to mention our Siamese twins as an example of either the EPR experiment or Bell's Inequality. This example is not intended to capture the richness of either and would hardly be recognized by an expert in modern physics. What it does do, however, is set up a framework to help us understand how they relate to causality.

Second Approximation

Einstein and his colleagues were unhappy with quantum mechanics, and they concocted a thought experiment to try to show that it was deficient in certain important ways. Instead of using twins, they asked us to consider two particles that interact with one another and then separate in different directions. Vaguely corresponding to Rule 1, we might take the conviction of all physicists that a quantity called the total momentum of a system is conserved. Just as with our twins, the sum of the momentum of the first and the second particle must equal a fixed value. Corresponding to Rule 2 we will take the equally strong conviction that no signal can go faster than the speed of light (approximately 186,000 miles per second). What we are now playing, of course, is Nature's own game as classical physics states it to be. These two laws of Nature have been analyzed both theoretically and experimentally to the satisfaction of physicists. When the particles are separated, let us disturb the momentum of one particle consistent with the results of quantum mechanics. The details get rather complicated at this point, but we can wonder whether the second particle will get the message to change its momentum instantaneously across space to keep the total momentum constant yet still not violate the law that we cannot send a signal faster than the speed of light.* To be more specific: If one particle of the two were on the sun and the other were on the earth, physics tells us that a signal traveling at the speed of light would take 8.5 minutes to go between them. If we change the momentum of the particle on earth, would the momentum of the particle on the sun change instantaneously or after 8.5 minutes? Einstein's strong conviction was that action could not take place at a distance and that we should prefer causality over the results of quantum mechanics. If I may paraphrase Einstein and brutally distort one of his statements to acknowledge our Siamese twins: On one supposition we should, in my [Einstein's] opinion, absolutely hold fast: the real factual situation of the system [in which the first twin is located] is independent of what is done with the system [in which the second twin is located], which is spatially separated from the former.

The objective of the EPR thought experiment was to show that quantum mechanics was flawed. If we insist that quantum mechanics is correct (and there is now ample reason to believe that it is), then the EPR experiment turns out instead to show that the separate particles are somehow intimately connected in a way that violates our usual ideas about causality. Bell's Inequality allows physicists to show *experimentally* that this is so. To quote the author of a book that will be cited later, John Gribbin (writing in about 1984),

> If he had lived to see it, he [Einstein] would certainly have been persuaded by the recent experimental tests of what is effectively a kind of EPR effect that he was wrong. Objective reality does *not* have any place in our fundamental description of the universe, but action at a distance, or acausality, does have such a place.

You might be curious about the experimental evidence that is so compelling. Let those who elect to remain with us consider it in the next approximation.

---

*The complicated details we have omitted include what *instantaneously* means in light of the special theory of relativity, how you can know the total momentum of the pair and the distance separating them in the light of quantum mechanics, and how the momentum of the first particle can be changed. The physics department of your local university can provide these details to the satisfaction of all.

Third Approximation

A thought experiment is one thing; an actual experiment is quite another. Most people find the heuristics that define an experimental solution to a problem more compelling than those that define a purely intellectual solution to one. This is usually the case with John Bell's formulation of Einstein's thought experiment, which allowed experimental evidence that action at a distance, acausality, or the rejection of local causes was ingrained in Nature.

Bell's original system in 1964 was not concerned with the conservation of momentum of two particles, as was the one that occupied Einstein. Instead it considered an equivalent system consisting of two protons and the conservation of a quantity called spin. As with our initial example of the Siamese twins, where numerical values were assigned to the raising and lowering of hands, two protons interacting in certain ways must conserve their spin: The sum of their spins must always equal the same value. The exact equations derived by Bell become more difficult and murkier to explain at this point for three reasons. First, the word *spin* is a technical word and does not have the precise meaning to a quantum mechanist as it does to the general public. Second, the particles do not change state in a simple fashion from $+1$ to -1 or vice versa, as did our twins. Since they are what are called $1/2$-spin particles, in fact, it takes two maneuvers to do so instead of only one. Third, the spin is what is called a vector quantity. Therefore instead of using ordinary mathematics to add the states as we did earlier, now we would have to use vector analysis. What Bell did was derive an inequality concerning the spin of protons using vector analysis and a branch of mathematics called set theory that must hold for the classical notion of causality to be true. If Bell's Inequality is violated, then the local realistic view of the world is false. Although subsequent experiments did, in fact, use protons to show that Bell's Inequality was violated, chronologically the first experiment (and most later ones) used still a third system consisting of photons and their polarization to demonstrate the same fact. Over the years a multitude of experiments with different systems have been performed, including every *aspect* of the problem, always with the same breathtaking results.

Do we live in privileged times? May we take the violation of Bell's Inequality as the current sota of physics, the results predicted by quantum mechanics in this area as true, the failure of classical causality as fact? Give me a minute.

I have just called the physics professor at The University of Texas* who is professionally most concerned with the matter for advice. His answer: Yes, we can. We live in privileged times. Bell's Inequality holds today.

Fourth Approximation

The next step for those who would like to pursue the matter would perhaps be to read the book by John Gribbin, *In Search of Schrödinger's Cat*, and the article by Bernard d'Espagnat, "The Quantum Theory and Reality" (*Scientific American* offprint #3066). Other popularizations of quantum mechanics certainly exist, but these references are particularly easy to understand. Dr. Richard Feynman, Nobel Laureate, has written a book, the *Feynman Lectures on Physics*, that may

---

*Prof. Arno Bohm.

appeal to some. Of course, if you have a lifetime to spare, you could begin a serious study of the field with Einstein, Podolsky, and Rosen, "Can Quantum-Mechanical Description of Physical Reality Be Considered Complete?" (*Physical Review* 47, 1935, pp. 777ff.).

Conclusions

Quantum mechanics is usually thought of as the physics of the small. But it has profound implications in the macroscopic domain as well. To quote Henry Stapp in an article in *Physical Review*:

> The important thing about Bell's Theorem is that it puts the dilemma posed by quantum phenomena clearly into the realm of macroscopic phenomena. . . . [It] shows that our ordinary ideas about the world are somehow profoundly deficient even on the macroscopic level.

Where is causality left in the world of quantum mechanics? I agree with the following quotation of the celebrated physicist Max Planck, when he writes (with my emphasis), "Causality is neither true or false. It is a most valuable *heuristic* principle to guide science in the direction of promising returns in ever progressive development," and with the equally famous physicist Niels Bohr, "[Quantum mechanics implies] the necessity of a final renunciation of the classical idea of causality and a radical revision of our attitude toward the problem of physical reality." Can there be any doubt as we read these comments that competent, knowledgeable, well-respected scientists deny that causality (at least as usually conceived) exists? In one of these quotations, causality is even identified as a heuristic, much to my delight.

Given the fact that both the philosopher and the physicist question causality, where does that leave us but with the

HEURISTIC: Causality.

By the way, note Gribbin's and Bohr's concern about physical reality in these quotations in addition to their concern for causality. This additional concern will serve as a convenient segue to the next section, in which we consider *consciousness* and *physical reality*. I will feel obligated to reproduce these citations again in a minute.

Consciousness as Consciousness

Be forewarned that I am now after a heuristic that is so radical that it makes even me hesitate. I refer to the

HEURISTIC: Consciousness

where the direct quotations of psychologists will be contrasted with the direct quotations of physicists, with each taking an unexpected role.

The direct quotation of experts is an effective heuristic for showing conflicting opinions. In most of the previous cases examined, the participants were from the same general intellectual discipline and, presumably, shared similar sotas. Contradiction is even more striking when it is supported by direct quotations of experts from widely diverse areas, where it is unlikely that the participants have a common background. In this situation the danger of taking a concept as absolute over which experts hold conflicting opinions is highlighted. In other words, we now have the

HEURISTIC: If experts in widely *different* disciplines disagree as to whether or not a concept is in question, perhaps we should be cautious.

First we give the microphone to one of the world's best-known psychologists, William James (with my emphasis): "I believe that *consciousness* is the name of a nonentity, and has no right to a place among first principles." Does this quotation leave any doubt about whether James thinks consciousness exists? His position is not an isolated one. It is supported by modern behaviorists such as B. F. Skinner, whose principal thesis is that "propositions that have been interpreted as referring to inner states of consciousness can be fully translated into statements about behavioral dispositions." *Self*, and I'm sure the behaviorist would include *consciousness*, is a device for representing a "functionally unified system of responses." So much for the point of view of influential psychologists on one side of the issue of consciousness.

I do not know whether I should include the physicists' concern over consciousness or not. If I appear to exaggerate at this point, I risk discrediting all that has come before. Still, if *all* is to mean "all" in *All i̲s h̲euristic*, I cannot afford to hesitate now. In addition, I have promised to give a quotation from a physicist accepting the validity of consciousness to pair with the quote by James denying it, and so I feel compelled to continue.

First Approximation

The promised quotation (again, with my emphasis) is by Eugene Wigner:

> The principal argument [against materialism] is that thought processes and *consciousness* are the primary concepts, that our knowledge of the external world is the content of our *consciousness* and the *consciousness*, therefore, cannot be denied. On the contrary, logically, the external world could be denied— though it is not very practical to do so.

Does this quotation leave any doubt where Wigner stands with respect to consciousness? It, and others of equally prominent physicists, can only be fully understood after we have developed some additional background.

Second Approximation

Again, it is a question of the results of quantum mechanics.

The traditional way to explain the strange role of consciousness to physics students who may not have the required mathematics or physics background has been to evoke a strange creature that has come to be known as "Schrödinger's cat." No argument is improved by exaggeration because the listener immediately becomes suspicious of what is to come. I do not know, therefore, if it is prudent to include the next example in support of my claim that *All i̲s h̲euristic*. It is so startling that not all scientists are agreed on what to make of it. Yet the physicist who put it forward is so influential and the discussions that have surrounded it have been so heated that I cannot avoid introducing this charming, but fantastic, feline to you.

In describing a thought experiment originally proposed by Schrödinger (the author of the fundamental equation that we included earlier in the text), I would like to quote extensively from Gribbin. This is not done because it would be hard to paraphrase the original article by Schrödinger (in *Naturwissenschaften* 23, 1935). The original example is both clear and concise. Instead I quote because the words of a competent scientist (John Gribbin, *In Search of Schrödinger's Cat*, 1984, pp. 203–204) writing for the general public assure us that the example is worthy of note and accurately described:

> Schrödinger suggested that we should imagine a box that contains a radioactive source, a detector that records the presence of radioactive particles (a Geiger counter, perhaps), a glass bottle containing a poison such as cyanide, and a live cat. The apparatus in the box is arranged so that the detector is switched on for just long enough so that there is a fifty-fifty chance that one of the atoms in the radioactive material will decay and that the detector will record a particle. If the detector does record such an event, then the glass container is crushed and the cat dies; if not, the cat lives. We have no way of knowing the outcome of this experiment until we open the box to look inside; radioactive decay occurs entirely by chance and is unpredictable except in a statistical sense. According to the strict Copenhagen interpretation [of quantum mechanics] . . . until we look inside, there is a radioactive sample that has both decayed and not decayed, a glass vessel of poison that is neither broken nor unbroken, and a cat that is both dead and alive, neither alive nor dead.

Let me sum up in one sentence: According to the Copenhagen interpretation of quantum mechanics, Schrödinger's cat is neither alive nor dead until *we* look into the box.

Third Approximation

With Schrödinger's cat (alive or dead) in the bag so to speak, let us look more technically at the relationship between consciousness and quantum mechanics. For those with some technical background, one paragraph should suffice to make the situation more precise. It may be safely omitted if desired.

Quantum mechanics is not concerned with a single, isolated system. Instead it describes a group of identical ones. The wave function ψ (which we met before) allows the prediction of the relative probability of all possible results that can be observed when one system is chosen from this group at random and a measurement is made on it. Assume that a specific measurement is made on just one of all of the systems from the group of systems. Further assume that the possible result of this measurement is one of a finite number of values. The wave function allows the advance calculation of the probability of getting each of these values. The wave function does not tell which result will be observed in a specific case, but it does give us the official betting line. Before we make the measurement, no individual system in the group has a definite value. The act of measuring causes one of the many possible worlds predicted by the wave function to be chosen and a definite discrete value to appear. The physicists refer to this effect as the reduction or collapse of the wave packet. The important notion is that the act of measuring forces the system to choose between the possible outcomes with a calculable relative probability. More precisely, the wave function permits us to foretell with what probabilities the system will make one or more impressions on us if we let it interact directly or indirectly with us.

With this paragraph in mind, let us pass the microphone back to Wigner:

> The modified wave function is . . . in general unpredictable before the impression gained at the interaction has entered our consciousness: it is the entering of an impression which we expect to receive in the future. It is at this point that the consciousness enters the theory unavoidably.

To establish the credibility of this view of consciousness and to show that others hold the same opinion, an additional quotation from an equally eminent physicist, John Wheeler, will help: "May the universe in some strange sense be 'brought into being' by the participation of those who participate?"

Physical Reality as Physical Reality

The previous section sought to demote consciousness to consciousness, but we seem to have gotten more than we bargained for. It appears that as a side effect of our efforts, we must also write

HEURISTIC: Physical reality

(objective reality, external world, or whatever), because the heuristics consciousness and physical reality appear to be so inextricably interwoven to the modern physicists that they must be added to our tapestry together. I will include another quotation, this one from Werner Heisenberg—"The conception of objective reality evaporates into the . . . mathematics that represent no longer the behavior of elementary particles but rather our knowledge of this behavior."—to secure this radical point and now repeat two quotations (the first by Gribbin; the second by Bohr) we have seen before, focusing attention not on causality this time, but on their view of physical reality:

> If he had lived to see it, he [Einstein] would certainly have been persuaded by the recent experimental tests of what is effectively a kind of EPR effect that he was wrong. Objective reality does *not* have any place in our fundamental description of the universe, but action at a distance, or acausality, does have such a place.

and "[Quantum mechanics implies] the necessity of a final renunciation of the classical idea of causality and a radical revision of our attitude toward the problem of physical reality." These comments certainly do make physical reality sound a lot like a heuristic.

Of course some of the other warp heuristics could just as well have been enlisted to make this point. We could have reviewed the longstanding clash between the realist (who believes that material objects exist external to us and independently of our sense experience) and the idealist (who believes that no such material objects or external realities exist apart from our knowledge of them). We could have considered the philosophical systems of logical positivism, Buddhist idealism, and that of Samkhya-Yoga, among many others. Or we could have selected relevant quotations from the writings of Berkeley, Hume, Malebranche, Leibnitz, Locke, Mill, Russell, and so forth.

The conflict in the direct quotations of James and Wigner, acknowledged experts in widely different fields, adds weight to the belief that the concept of consciousness and physical reality suffer, at worse, elimination as an absolute characteristic of our universe and, at best, contradiction and redefinition. In fairness I should mention that some experts in psychology and physics might not agree with the quotations attributed here to spokesmen for their fields—but then disagreement among experts in the same field is by now a familiar heuristic.

We are forced, at last, to agree with the words of the German philosopher Friedrich von Schiller (with my emphasis):

> If we accept the notion that the external world is strictly unknowable, then the status of our hypothesis about the world is that of fictions, inventions, and *heuristic* devices to answer the questions to which we need answers in order to live; but the validity of these hypotheses will depend upon human criteria of factualness, logic, and plausibility and not upon their agreement with extrinsic nature.

As spooky as an analysis of these transitions seem to be, I believe I will just write *consciousness* and *physical reality* to be on the safe side.

A quick glance at the original list of heuristics shows that we are making good progress. Only two woof heuristics remain to be considered, but they are some of the most intractable ones with which we must deal.

Science as Science

By most accounts, science strives to give a better and better account of the external world, of Nature behind the veil, of what really is. Figure 35* illustrates this goal by comparing an overall set U with several smaller subsets.

As before, U is taken to include all of the truths of Nature or, if you prefer, all of the questions that can be asked of him. Again, the crosshatched subsets, A, B, B', C, D, and I represent specific theories, principles, and laws that enclose those questions for which each provides the answer. At times more than one scientific theory will be needed to explain some facts of nature and the appropriate sets will overlap, as shown with C and D. No one claims that we have all of the answers to Nature's questions at the present time. Therefore, the sum of the crosshatched areas does not completely cover U, but only a small, the Chinese philosopher would say insignificant, part of it. The scientist's job is to increase the coverage.

The influence of science is so great, the results achieved with its aid so far-reaching, and its hold over our thoughts so profound that it is hard not to accept science as an absolute way of knowing about our world. There can be little dispute that the scientist is an expert in examining our world or that his heuristic for doing so is a valuable one. Typically a scientist has dedicated his life to the pursuit of his version of truth, and this truth has left its mark on our lives, worldview, and grammar. We can neither ask *Why?* nor answer *Because*

---

*This is essentially a redrawing of Figure 9 with a few minor modifications.

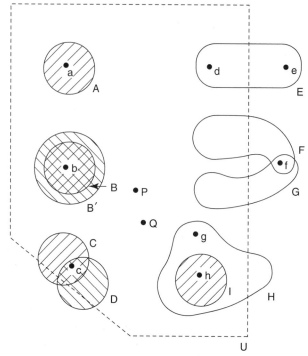

Figure 35 World of Science

without admitting our indebtedness to him. Often highly intelligent, articulate, well trained, and imbued with a self-confidence that infects all around him, it is hard not to believe his claim of absolute knowledge. Upon examination, however, we find that the scientist just presides over one more heuristic, valuable though it may be.

Toppling science from its lofty position will not be easy, but I feel obligated to make the effort. This section looks closely at science from the following five perspectives to establish that it is a heuristic. It considers science's dependency on other heuristics for its operation, its incompleteness at the present moment, how it came into being, and its current status. These four considerations will be followed by quotations of eminent scientists casting doubt on science as an absolute way to achieve true knowledge.

Dependency on Other Heuristics

The most incisive demonstration that science is a heuristic is its complete dependency on other heuristics for its operation. We have already examined many of the essential concepts it requires in some detail. Where would science be without induction, arithmetic, mathematics, deduction, logic, truth, causality, and physical reality? If it cannot add, deduce, and, as we will soon see, perceive with certainty, how can it ever claim to give us a certain picture of the world? I can find no escape from the conclusion that science must be a heuristic if it depends so fundamentally on other concepts that are clearly heuristics.

Although science's track record is good, it is far from perfect. To allay any doubts that it just might be wrong in any specific instance, we will need additional heuristics to sort out when science provides a correct answer and when it does not. The exact location of the dashed boundary of U is unknown; at times, the best theories turn out to be in error. For example, in certain situations, even Einstein's theory of relativity is simply wrong. It predicts its own demise because it includes mathematical objects known as singular points. Since the present sota of science forbids singularities in Nature, the crosshatched set that represents relativity must extend beyond the dashed line, as indicated by the set C or D in Figure 35. With no way to determine the precise location of the dashed line, knowing for sure when a theory is correct and when it is not depends on additional heuristics. On this basis alone, the theories of science represented by A, B, B', C, D, and I are beginning to look a lot like heuristics themselves.

Science's claim of providing sure knowledge or at least of converging asymptotically to sure knowledge depends on still other heuristics. One of these has already been eluded to earlier, but it bears repeating here. To replace one theory, B, with another more acceptable one, B', science needs decision rules (heuristics) to define what is meant by "more acceptable." Science must decide when Einstein's theory is to replace Newton's theory and needs help to do so. Nearly universally accepted is the rule that a new theory must be simpler (Occam's razor) or more comprehensive than the one it replaces.

Occam's razor,* often called the principle of parsimony, implies the methodological principle of economy in explanation. It is often stated as "entities are not to be multiplied without necessity." The scientist, to justify a preference for the simpler of two alternate explanations, uses this heuristic.

Here, once again, we have an example of the "less clear being used to enlighten the more clear." The philosophical notions of *simplicity* and *comprehensiveness* are neither simpler nor more comprehensive than science and must be taken as heuristics themselves. Indeed, we enter a vicious circle if we try to establish scientifically that simplicity is the correct way to proceed. On the other hand, if we go outside of science to try to do so, where does that leave science's supposedly fundamental nature? Substituting some of the more exotic decision heuristics such as "ease of falsification" for simplicity and comprehensiveness as is sometimes suggested does not solve this problem. Science's claim of providing sure knowledge must falter because of its dependency on heuristics to know how to proceed. As is all too often the case, not only the cathedral but also the flying buttresses that support it rest on sand. If everything is a heuristic, then this situation is not just a problem of science but one of all intellectual architecture.

Ironically, Occam's razor supports the basic claim that everything is a heuristic. If the scientist can use it, then so may I. On the basis of the previous discussion it may not yet be safe to conclude that everything is a heuristic, but surely it is safe to conclude that some things are. As a result, the proposition "there is at least one thing that is doubtful" is easier to endorse than the proposition "there is one thing that is certain." If you do not agree, now is the time to produce your candidate for absolute certainty that is more compelling than *any* heuristic I may produce to doubt, and I will sic the nearest skeptic on it. Even

---

*There is no evidence that William of Occam ever used this formulation.

in generalization the claim that "everything is open to doubt" (the position of the skeptic) seems more reasonable than the claim "everything is certain" (to my knowledge, the position of no one). I grant, of course, that most people's sotas probably contain a mixture of doubt and certainty. But of the two extremes, universal doubt does seem to have a clear advantage over universal certainty.

Now apply Occam's razor in the form "plurality is not to be assumed without necessity." Moving from "everything is in doubt" to "some things are in doubt and some things are certain" demands the additional heuristics certainty, dichotomy, classification, and maybe even logic and language. Instantly the world becomes less simple and far more complicated. Can this increase in complexity be justified? Occam continues. He considers as a pseudo-explanation any entity the necessity of which was not specifically established by "evident evidence, evident reasoning, or required by articles of faith." To which of these three categories are the additional heuristics needed to pass from universal doubt to a mixture of doubt and certainty to be assigned? If science can insist on Occam's razor as one of its guiding principles, so can we. Happily it supports the claim that *All is heuristic.*

Science is nothing without heuristics. It depends fundamentally on other heuristics; it needs heuristics to know if its pronouncements are true or false; and it requires heuristics to make progress. Should we not consider science a heuristic itself?

Incomplete

I first alluded to my next reason for suspecting that science should be considered as a heuristic near the end of the section giving the signatures of a heuristic. I amplify that discussion because of its importance. The scientist is constrained to live within the *crosshatched* region in Figure 35. This is his sota, and it contains the only knowledge to which even he claims access. He must take it on faith that a later, more comprehensive sota will not rob his current one of all significance. What confidence does he have and how is it buttressed that an unknown fact, call it P, outside of what is now known will not eventually show that his current sota lies completely outside of U and is, therefore, completely false? Even if he could be assured that such a P does not exist, how can he be certain that another fact, Q, does not exist that is capable of blocking the essentially monotonic expansion of the crosshatched area, much as Gödel's Proof did for the sota mathematics? Given the consensus that Nature is subtle, complex, and only partially understood at present, combined with the admittedly small area of U that is presently covered by science, it takes an extremely fundamental and daring assumption to do away with the possibility of a P or Q—one prudent scientists should be loath to make. The incomplete picture science gives of the world at present reduces any claim to certain knowledge to one of blind faith. That is, unless we freely admit that science is a heuristic.

Origin

Science, using the word in anywhere near its present connotation, is a human invention. We find it strange to think of other living things (even the higher primates) using science in any developed sense. Contrast science with perception as a way of gaining knowledge. Is it conceivable that cockroaches contemplate the nature of the universe, develop theories about it, and test them? I think not.

Yet, cockroaches certainly seem to sense their world and act on that knowledge. Science seems so artificial and man-made; perception, so much more natural. Science is undeniably a product of the human mind.

Not only is science a human invention, it is a relatively new human invention. Many historians credit the Ionian natural philosophers of the sixth century B.C. as its founders. When Thales taught that everything was made of water and his disciple, Anaximander, disagreed, the human species saw the birth of what has variously been called the Greek way of thinking, the comprehensibility assumption, and the scientific myth. Not surprisingly, I call it the Greek heuristic. It is instructive to remember that those cultures derived from ancient China and informally based upon what might be called the "Chinese way of thinking" do not endorse this heuristic. Other cultures, such as the French, whose once strong rational tradition is now greatly weakened by modern philosophy under the influence of Martin Heidegger and, perhaps to a certain extent, Henri Bergson, are retreating from it. To be sure, some historians have found traces of doubt and criticism before the sixth century B.C. Others have found Asiatic anticipations and variants of this same approach to explanation in Indian and Chinese literature. But the sages of the Milesian school in Ionia first made the definitive hypothesis that the world was comprehensible and accessible to critical analysis. This assumption is therefore a Greek invention, one of the pivotal inventions in human history. So pervasive has this hypothesis become that the historian, Edith Hamilton in *The Greek Way,* has been led to assert, "The Greeks were the first scientists and all science goes back to them." And another historian, John Burnet in *Early Greek Philosophy,* concludes, "[I]t is an adequate description of science to say that it is thinking of the world in the Greek way."

This new, Greek, heuristic can hardly claim to have stood the test of time. The cockroach has been around for millions of years; science, as we have just seen, at best, only since the sixth century B.C. Because science is only a recent human invention, we must admit that other more advanced inventions are possible, even perhaps that they are probable. It will be sad if the human species becomes overly fascinated by its new heuristic *science*, assumes it is infallible, and chases it to oblivion.

Present Status

A fourth aspect of the claim that science provides sure knowledge is worth noting. Most scientists will agree that the earlier representation of a scientific theory as being either right or wrong is naïve. Actually, a theory does not overlap the boundary of U: at times lying within, at times lying without. Instead, many scientists admit that all scientific theories are approximations, hypotheses, guesses, conjectures, and heuristics always subject to recall. None is true anywhere; all are false everywhere. Some are just falser than others are and the scientist has a preferred taste for the false over the falser. In this view, Einstein's theory of relativity does not claim to be absolutely true, but it does claim to be a better approximation to the correct relationship between two masses than is Newton's law of gravitation. Figure 36, a three-dimensional view of the set U, will help make this clear. Now the subsets or theories are located at different distances above the plane of U corresponding to the relative validity of each as an approximation. The nearer it is to the surface of U, the better it approximates the real world. From the point of view of the scientist, science gives us a cloud

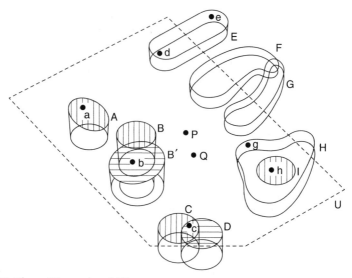

Figure 36 Three-Dimensional View

of heuristics hovering over the facts of the real world to which they correspond. In this traditional view, as science makes progress, this cloud both expands and descends.

Unfortunately this cloud holds some surprises. A one-to-one correspondence between a point in U and the best available theory does not necessarily exist. For instance, consider the point in U that represents the *true nature of light* and look above it for the corresponding *current best theory* in the cloud. What we see are two different, conflicting theories depending on which cloud we choose to consider. Since by one theory, amply supported by experimental evidence, light behaves as a wave, and by another theory, likewise based on experimental evidence, light behaves as a particle, which cloud is to give the current best theory of light? To explain this odd state of affairs, Niels Bohr introduced the Principle of Complementarity into modern physics. He recognized that both of these theories of light were self-consistent and corresponded to experiment and that contradiction only appeared (1) when an effort was made to combine both the wave nature and the particle nature of light within one theory or (2) when an effort was made to design an experiment to exhibit this dual nature of light. As the physicist Wolfgang Pauli explains, "I can choose to observe one experimental setup, A, and ruin B, or choose to observe B and ruin A. I cannot choose not to ruin one of them"; or as another physicist, Heisenberg, insists, "What we observe is not Nature itself, but Nature exposed to our method of questioning."

The Principle of Complementarity places unexpected limits on our ability to achieve a definitive, unique match between the cloud of heuristics and the real world.

Direct Quotations

Not all scientists, to be sure, are confident of science's ability to achieve sure knowledge. This is the final reason why we should consider science as a heuristic. Bronowski, for one, reminds us that theoretical science is an attempt to uncover an ultimate and comprehensive set of axioms, and that in the light of Gödel's work this is impossible. Wheeler, after concluding that "in the end there is no law except that there is no law," compared his release [from physical law] to a bath of cold water—"at first it's a shock, then it's immensely stimulating." Jeans compared his release to a release from prison. Bridgeman admitted that he found science dissolving in his hands and the universe and existence without meaning. Others, with noticeable regret, have concluded that science is a myth. Still others have agreed that it ultimately rests on faith. And others that no one really knows what it is about—only that it seems to work. Is it not, perhaps, just one more heuristic?

Conclusions

Based on science's dependency on known heuristics, its present incomplete knowledge of the world, its origin, and its present status coupled with what some of the most thoughtful scientists have said about it, we are now in a position to answer some important questions that are often asked about the whole scientific enterprise.

What Is Science?   Some of the wisest humans have tried to answer that question and failed. "Science is the pursuit of truth"—but what is truth? "Science progresses by self-correction"—what is progress? "Science gives us knowledge of reality or the external world"—is there a dispute over the nature of reality and the external world? "Science uses mathematics to build on a foundation of definitions"—can we be sure of mathematics and definition?

In view of Figure 36, the cloud that represents the best model of reality consists of heuristics; the changes and improvements in this cloud over time require heuristics; the mapping from reality to cloud is at times uncertain and, therefore, heuristic; and as we have already seen, the external world is also a heuristic and therefore must become a part of the cloud of heuristics itself. Only a cloud of heuristics exists. Everything once thought absolute, including the concept of the absolute, deserts the plane of the absolute to float suspended with the rest as points in the cloud. This accumulated uncertainty forces us to identify science as the

HEURISTIC: Science.

Engineers do not just use science as a heuristic, as we insisted on page 86; it is one. But let us not become overly disturbed by this fact, for in a later section science, like the Phoenix, will rise to assume its proper importance once again if only we will let the ashes get cold.

What Is a Scientific Argument?   Collecting and repeating my earlier sentiments in one place: Any argument based on induction, language, arithmetic, mathematics, deduction, logic, or truth must be considered as a heuristic argument. Now at last I am forced to insist, any argument claiming to be scientific must also be taken as a heuristic argument.

What Is the Scientific Method?   Perfectly parallel with the final definition of the engineering method we achieved earlier, we now have the new definition of the scientific method as the

HEURISTIC: Use scientific heuristics.

What Then Is the Scientific Enterprise? I am afraid I am at last forced to disagree with Von Kármán, who was quoted on page 13. Scientists do not discover what is. At the best, they discover what *is*. Science is applied engineering; the heuristics it uses are objective reality, time, causality, truth, mathematics, logic, comprehensibility, ...

Perception as Perception

"Stop, I've had enough," I can almost hear my engineering colleagues complain. "Your words nearly hypnotized me into believing that mathematics was flawed, that science betrayed, that logic could not be trusted, that reality was dissolving in my hands, that consciousness, truth, causality were all gone. They might mesmerize the philosopher into believing these things, but they don't fool me."

"Do you see that bridge over there?" my colleague continues. "I designed it. Let the liberal arts student believe such nonsense. I will have no part of it. I *see* my bridge, I *feel* my bridge, my bridge exists."

How, the engineer asks, can anyone seriously believe that his own hand does not exist, that a rock would cease to exist if a mind were not perceiving it, or that motion is impossible? The engineer sides with G. E. Moore and raises his hand, with Dr. Johnson and kicks the nearest stone, and with Diogenes the Cynic and gets up to walk around.

"I sincerely believe in my bridge," my colleague says with finality. "I cannot think otherwise."

The next warp

HEURISTIC: Sincerity of belief and the inability to disbelieve are poor justifications for claiming that a belief is true

acknowledges the power of this feeling.

The dubiousness of some of the most sincerely held beliefs is so frequently demonstrated that it needs little amplification. Can we doubt that the Pope profoundly, intellectually, emotionally, even dogmatically, believes in Christianity? Can we doubt that the Zen Master profoundly, intellectually, emotionally, even dogmatically, believes in Buddhism? Can we doubt that some of the most learned philosophers of the Western world profoundly, intellectually, emotionally, even dogmatically, believe in atheism? What more sincere acts of immolation do you require than those of the Christians who died in the forum, the monks who have doused themselves with gasoline and set it on fire, and those who put their soul in peril for an idea? In the Western tradition these views are irreconcilable. Someone is simply wrong. Someone's sincere belief is not true. The same analysis applies to the alliances to widely divergent philosophical and governmental systems. Can we doubt that the dogmatist is truly dogmatic and the skeptic is truly skeptical? Are we to deny that the Kamikaze, modern-day terrorists, and the Assassins were sincere in their disagreements with other people and other forms of government? Except in a heuristic world, where belief is belief, wrong is wrong, and contradiction is contradiction, these various examples of mutually exclusive, sincerely held beliefs cannot coexist.

We do not have to go to these persons who flaunt their sotas to find sincere beliefs that turn out to be wrong. Each of us, at one time or another, has profoundly, intellectually, emotionally, even dogmatically, held a belief that we later recanted. Were you an atheist, but now a theist, or the contrary; were you once for war, but now against it, or vice versa; were you dogmatic about a particular subject, but now more skeptical, or the reverse? Is your view of time, of arithmetic, of logic, the same as it was when we began our discussion? No one's sota has been without change, change even in those areas seemingly the most sure. Belief, no matter how sincerely held, no matter how compelling, no matter how certain, is a poor justification for truth. Believing something based on the *strongest personal terms* is risky business.

This heuristic does not just set up a strawman for attack. People do believe just because they cannot do otherwise. Look at the book in your hand. Why else do you probably still believe that it absolutely exists despite my best efforts? No doubt you cannot even conceive of there not being a book in your hand. "Seeing is believing," you say, but *seeing* is precisely the issue. You must agree that your confidence in this world is based in large measure, if not exclusively, on the testimony of your senses, on how you perceive the world. Accepting perception as a heuristic is our final challenge.

The scholars who study the faithfulness of the senses as they act as ambassadors of the external world remind me of witches at a witches' Sabbath vying with each other to produce the most powerful potion to cloud our senses. These sorcerers include some of the most famous philosophers, psychologists, and physiologists from the time of Democritus. Plato, Hume, Berkeley, Descartes, Kant, and Locke immediately come to mind, but their intellectual progeny still meet at the full moon. No two covens use the same recipe to concoct their brews—brews that bear odd-sounding names like the commonsense, the direct realism, the critical realism, the representative or causal, the sense-datum, the Gestalt, and the phenomenalistic theories of perception. Instead of the expected eye of newt, toe of frog, or tongue of dog, the caldron around which this group dances contains some pretty strange stuff. In the bubbling waters, we find elliptical pennies, crooked oars, ambiguous duck-rabbits, Macbeth's dagger, and phantom limbs. All this effort has been to produce a powerful antidote to our confidence in perception's ability to correctly tell us of the external world by evoking illusions, hallucinations, dreams, and at least one very malicious demon. Using my own brew, I'll personally add hypnotism to this list.

Before we get too involved in a discussion of perception, we must all agree with one stark fact: At times, the senses simply lie. Figure 37 is a classic optical illusion that forcibly makes this point. Please look at the circles in this figure. Which of the two center circles is the largest, or are they the same size? Now measure them and look again. Try as you may, nothing you can do will make what you see conform to what you measure.

This optical illusion is not an isolated example. Look at Figure 38. Which of the three lines x, y, and z is an extension of w? With a ruler or the edge of a sheet of paper, check your guess. Once you know the answer, agreement between what you see and what you measure is somewhat easier to achieve than in the previous case, but it is still hard to do. This last illusion is the basis of some of the large-scale magic tricks we see on television in which large objects or land masses such as Diamond Head in Hawaii are made to appear to move or disappear right before our eyes. A professional magician could easily make an engineer's bridge disappear or move in a similar fashion. The point is that our senses undeniably lie, and that fact is as certain as the certainty that you now hold a book in your hands. We may argue about why, when, or how the senses

Figure 37 Circle Size Illusion

lie, but these simple, classic optical illusions definitively show that lie they do. Once they have been shown to be liars, the burden of proof shifts to them to show that they are not lying in a specific instance.

A complete survey of the literature of perception is impractical because of the wealth of research available. This fact alone should raise our suspicions that perception is a heuristic as it has done with many other concepts in the past. Since the arguments of the ancient skeptics still have the power to cast their spells and cloud our senses, a representative sample of their work will bring some order to our investigations and prove that there is more to perception than meets the eye. Then this section will examine some of the more modern skeptical results such as the arguments based on the condition known as synesthesia, illusions,

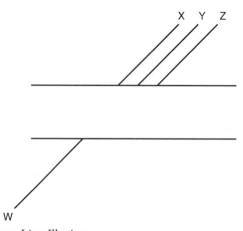

Figure 38 Continuous Line Illusion

dreams, the work of a malicious demon, and hypnosis. I am confident that at last we will be forced to accept that perhaps you are not holding a book in your hand, but only dealing with a very good and a very persuasive

HEURISTIC: Perception.

Arguments of Aenesidemus

Certainly the most famous attack on perception in the ancient world is by the skeptic Aenesidemus. In his first four skeptical arguments (tropes, modes, or positions) out of a set of ten, he argues that the senses are not to be trusted. Aenesidemus gives these reasons: Animals perceive things differently, different humans perceive things differently, man's senses perceive the same object in various ways, and man's circumstances alter what he perceives. They are sufficiently important to be considered briefly now, leaving the task of treating each in greater detail to Sextus Empiricus, whose version, I am authoritatively informed, is the longest and the most convincing.

Species Differ   Different species of animals have different sense organs and, hence, view the world in different ways. For example, the fish that belong to the family *Mormyridae* release impulses that set up a system of electric fields. They have sense organs along the sides of the body that can detect changes in these electric fields and in this way the mormyrids obtain an accurate image of the objects in their environment. Other animals have senses that are absent in the human or use those held in common with humans more acutely. Bats sense ultrahigh-frequency sounds although humans do not; rattlesnakes sense infrared radiation unlike the human; and dogs sense smells that cannot be detected by members of the human species. How can we ever be certain that humans, not some other animal, have the requisite senses to correctly perceive the world? Montaigne has reminded us that for all we know, "We may be as far removed from accurately perceiving Nature as a blind man is from seeing colors."

The necessary senses to obtain an accurate picture of the real world need not be limited to those senses that have already been found in animals. As yet unknown, but perhaps essential, senses are quite possibly still to be discovered or developed. What we think we now see is largely due to the sensitivity of the human eye to the visual portion of the electromagnetic spectrum between four hundred and seven hundred nanometers. Recall that quantum mechanics represents Nature in terms of Ψ and $\int\Psi\Psi^*d\tau$. If future humans develop a way to sense either of these two quantities directly, the world may well appear as a continuum instead of being divided into rabbits, water, and humans, as it currently seems. This is irresponsible speculation to be sure, but it would be the ultimate irony if we were all parts of the same stuff currently singled out as individuals only because of present, but nonessential, limitations on our senses. Even were we to have the thousand senses of Micromégas, we could never be absolutely sure that we had all of the senses necessary for a true knowledge of Nature.

Humans Differ   No two humans perceive the world in precisely the same way. This casts doubts on any one person's claim of correctly perceiving the world. Recognition of this fact is the second skeptical argument against the fidelity of the senses recognized by Aenesidemus. Several examples demonstrate the power of this trope to discredit the senses. As one simple example, figure C on page 1 of the Insert shows the familiar test to detect color blindness. What numbers do you see? That other people see completely different numbers should be troubling.

Disagreement over what different people perceive is not limited to sight. Humans also differ greatly in the sense of smell. The largest scientific experiment ever performed was conducted by The National Geographic Society in 1986 to make this point. Over 1.5 million people from all over the world "scratched and sniffed" at six small samples of scent and returned survey forms. The results definitively demonstrated that a wide variation in the sense of smell exists between people. This experiment concluded, among other things, that women smell more acutely than men, that reactions to odors vary widely around the world, that two persons in three have suffered a temporary loss of smell, and that 1.2 percent of those returning survey forms cannot smell at all. As to the sense of taste, only 70 percent of the population report that they taste the chemical phenylthio-carbamide as bitter when it is presented to them on a small piece of filter paper. Finally, when our children were young, my wife could hear our baby crying in the middle of the night when I heard nothing. Research seems to substantiate the unusual ability of a mother to hear her baby cry. Whether in the case of sight, smell, taste, or sound, humans simply differ in the way they perceive the world.

All of the examples cited are convincing because they appeal to our concrete experience. We can look at the numbers on the test for color blindness, we can smell the scents for ourselves, we can taste phenylthio-carbamide and judge whether or not it tastes unpleasant, and we can hear a baby cry. In the world of religion, claims of differing perceptions of the world exist that are not so direct or confirmable. As one specific example, the Zen master reports that he perceives the world as a whole rather than differentiated into parts. Similar claims exist in other religions of the world. Because these claims appear less scientific, we cannot confirm them here, but since they are reported by some of the most sincere, presumably honest, people we know, we must admit the possibility that these differences in perception between different people exist. Since different humans perceive the world differently, one must be extremely egocentric to presume that his way of perceiving it is necessarily correct.

Senses Differ   On the list of the traditional skeptical arguments recommending doubt about the senses, the third trope is based on the different structures of the sense organs. If the Crab Nebula is viewed by a variety of apparatus, one sensitive to radio waves, another to x-rays, and a third to visible light, the nebula appears so different that the uninitiated would never identify the separate images as coming from the same object. In a like manner, since the human sensing devices differ as much as do the eye, ear, tongue, and finger, when they disagree it is impossible to tell for sure if they are reporting a single reality or not.

The examples of this trope known at the time of Aenesidemus are somewhat unconvincing, but more modern examples have been created to show the full strength of this argument. The ancient skeptics were concerned that the eye saw a three-dimensional object in a painting, but when the canvas was touched, the finger reported it as flat. This perplexity is only heightened by the invention of perspective in art. Paintings are now so commonplace that it is unlikely that anyone would be confused by this effect, although modern trompe l'oeil paintings still possess the power to deceive. We do not know when Narcissus first looked at his image in the stream and fell in love, but we personally feel the power of the reflected image in the mirrors of today to make us believe in its existence. Technically a mirror creates what is called a virtual image on the other side of the glass. This image is fully integrated into our world and, with the exception of a reversal, appears indistinguishable from the original. Reach out to touch the image watching you as you comb your hair and your clone does likewise to block your entry into the mirror's secret world. We tend to forget this common contradiction between the senses of sight and touch because we are so familiar with it, but it has left us with the cliché "It's all done with mirrors" to explain any trick of the senses we do not understand.

A modern variant of this theme is far less well known and still has the power to amaze. This new illusion consists of two parabolic mirrors assembled similar to the two shells of a clam with the reflecting surfaces on the inside. When a small object such as a thumbtack is placed on the inside of the bottom shell and viewed though a hole cut in the top shell, it appears to float above the device. What is unusual about this effect is that instead of the virtual image of the conventional mirror, this device creates a positive image suspended in the air for all to see. Once again, this image is indistinguishable from the real thumbtack. It appears solid. It casts shadows. It may be viewed from all sides. It is fully integrated into our world. When we reach out to touch it, however, we are not blocked as in the previous case. Instead of pricking our finger, however, we feel nothing and are left with thin air. Consider several even more perplexing examples. We are all familiar with the Polaroid glasses or those with red and blue lenses that permit us to see movies and comics in three dimensions. We are not overly concerned by this contradiction of our senses because we have an intervening pair of glasses. We know it to be a trick of the senses, but then a trick of the senses is precisely the point I am trying to make. Another device that gives us ambiguous information from different senses is the hologram. A hologram casts a three-dimensional image that has parallax and, once again, is completely integrated into veridical experience. In some cases a person may walk around a hologram and see it from all sides. When the hand reaches out to touch the object, it is left grasping at thin air.

With mirrors, special glasses, or even holograms, we feel tricked by a scientific apparatus. Although each makes the point that different senses give conflicting information about the world, somehow we feel that the deception is unfair. One last example is so simple, so compelling, and so straightforward that no one can complain that an unfair advantage is being taken of the senses. To understand this last example, we must first become familiar with the optical illusion in Figure 39. This cube can appear in two orientations: one with the corner marked x projecting toward us and the other with it pointing away from us. Some people can see the difference easier by considering the top plane of the cube. At times we are looking at it from above; at times, from below. Once both views can be perceived, most people are able to make the orientation of the cube flip between the two possibilities at will. If a three-dimensional version of this cube constructed from very fine wire is held in the hand, many are able to flip the cube into the opposite orientation from the one they hold. This effect is easier to achieve if the wire is painted a brilliant, florescent orange and viewed in a darkened room under ultraviolet light to minimize shadows that give away the correct orientation. The effect of the eyes giving one orientation and the hands giving another is very unsettling. If the senses disagree as to the nature of the external world, how can we assume that only a single external world exists?

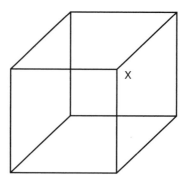

Figure 39 Necker Cube

Same Senses Differ   Finally, even if the human possesses all of the senses needed for an accurate view of the world, even if all humans could agree on what the senses report, and even if the individual senses gave a single coherent view of the external world, an additional problem must be solved. The world appears different to the same sense in different circumstances and at different times in our lives. No one denies that the taste of a substance is affected by the health and age of an individual, that the sense of hearing becomes more acute when one is blinded, or that the world appears different when one becomes intoxicated. An additional result of the smell test described earlier was that a woman's sense of smell becomes more acute when she is pregnant. A single sense does not give the same view of the world at all times.

Experience matters. What we see, what we taste, and what we hear depends the experiences we have had. Reconsider an example that was used earlier when we needed to replace *language* with *language*. At that time, I reported on an informal test I had performed with generations of students to see if engineering and liberal arts students were as good at distinguishing colors as were trained artists. We used Plate 1 to conclude that previous experience with color matters. As to the sense of taste, we are all aware that some people are better than others at identifying the complex flavors in wine, and that with practice we can become more proficient ourselves. If it is true that people can be trained to detect differences in colors and differences in flavors, we are forced to admit the same sense perceives the world differently based on experience.

A classical optical illusion shown in Figure 40 is included to allow you to experience for yourself that our experiences matter in what we sense. In the figure, do you see a young or an old lady?* If you have never encountered this illusion before, you will probably see either one or the other, but not both. With

---

*This well-known illusion once appeared on the cover of *Science* magazine. To see the young lady, note: She is looking away from us, we see her left ear near the exact center of the picture, and the small black stroke at the bottom of the picture is her necklace. To see the old lady, note: She is in severe profile with her chin below the picture looking out of a fur coat, the young lady's ear has become the eye of the old lady, the chin of the young lady has become the nose of the old lady, and the black line is now the old lady's mouth.

Figure 40 Old-Young Lady Illusion

patience and experience, the figure begins to alternate between the two. If you are so sure of the reliability of your senses, how do you explain the difference in what you see now and what you saw a few minutes ago?

When we began the discussion of the same sense seeing different things because of different circumstances, we poked gentle fun at the ancient skeptics for using paintings as an example. Let us end the discussion by poking gentle fun at ourselves. On the very low ceiling of the lobby of Rockefeller Center in New York City, we find a painting that deceives us even now. This power is explained in the Rockefeller Center Visitor's Guide available in 2002 as follows: "Looking up, you'll see another Sert mural, called *Time*, depicting the past, present, and future. The three trompe l'oeil figures stand on columns, and seem to shift their weights as you observe them from different areas of the lobby." Unfortunately I cannot capture the effect for you on paper and no photographs are available from the archivist, but I give my personal assurance that you will not, cannot, believe your eyes.

Given the wealth of examples we have just seen, how can we ever be sure that no circumstantial conditions are now in effect that fundamentally distort our perception of the real world or that we do not lack the experience to correctly use our senses?

These four tropes are not the only ones on Aenesidemus' list of ten that deal with perception. Nevertheless, they do indicate the skeptic's early preoccupation with and success in challenging the reliability of the senses.

Since Aenesidemus, the attack on the veracity of the senses has been unrelenting. Three of the main volleys have been the argument from illusion (with hallucinations as a special case), the dream hypothesis, and the assumption of a malicious demon. To these three I will add synesthesia and my own conviction that hypnotism casts the ultimate doubts on perception.

Synesthesia

"A man who tastes shapes," "A pain that is orange," "The color red that rings louder in your eye," "The taste of blue that lingers on your fingertips"—are we awash in metaphors? Are we dealing with people who are hallucinating? Are we still speaking conventional English? Reach for the nearest dictionary:

> Syn·es·the·sia: A condition in which one type of stimulation evokes the sensation of another, as when hearing a sound produces the visualizations of a color.

We are not dealing with figures of speech, but with a real, human phenomenon susceptible to research.

To this point in our analysis of perception, we have been concerned with the senses taken individually. We commented on errors in sight, smell, and taste and drew conclusions because of their unreliability. Now we must consider an odd condition in otherwise normal people for whom two or more senses are comingled. Researchers estimate the incidence of synesthesia as one in twenty-five thousand individuals worldwide, including some famous people such as Russian novelist Vladimir Nabokov and British painter David Hockney.

At the very least, synesthesia corroborates the earlier claim that different people perceive the world differently, reawakening all of the concerns expressed earlier. In addition, research suggests that it is hereditary, probably produced by a dominate gene on the X chromosome. People who experience synesthesia are quite happy with their world. In some cases, they actually have a competitive advantage over others and might well be selected by evolution. For example, memorizing names and addresses is easier because the colors, sounds, or tactile sensations provide additional clues and a richer information environment. To quote Shabana Tajwar, a synesthete, when trying to remember a name, "I knew the name was green. It started with F and F is green." If this turns out to be the case, the way the majority of us now conceive of the world could someday be a minority opinion.

Argument from Illusion

The argument from illusion just picks up the baton of the early skeptics and continues the relay into the present day with the invention of even more convincing illusions. As was stated and demonstrated at the beginning of the present considerations, people are demonstrably deceived by their senses—that's all there is to it. Whether we consider

1. The drugs that make us see double images,
2. Optical illusions such as those that make equal lines appear unequal,
3. The famous distorted room that makes a grown man appear smaller than a boy,
4. The ambiguous picture that at one moment appear as a duck and then at the next as a rabbit,
5. Mirages,
6. The appearance of motion in the cinema,
7. The phantom limbs soldiers who have had their arms or legs amputated report,

8. The penny that is taken as round although it appears elliptical when seen at an angle, or

9. The stroboscopic effect that makes complex moving machinery appear to stop so that it may be inspected for imperfections,

we can only become more convinced that the senses simply lie. To these examples should be added documented cases of hallucinations in which physical objects are positively perceived when none are present. Typically these objects are well integrated into veridical experience in the sense that they cast shadows, they vary in size as they are approached, and their presence is rationalized by the percipient. Because of this complete integration, a person can never be sure when he is being deceived and must accept the possibility that he is being deceived at any given moment. Whether through illusion or hallucinations, we have ample evidence to question the reliability of the senses.

Argument from Dreams

Another skeptical argument that has appeared from time to time, principally in the work of Cicero, Montaigne, and Descartes, is the dream hypothesis. Although admittedly some modern philosophers feel that they have laid the phantom world of dreams to rest, this hypothesis can be made very persuasive and has a habit of returning from the dead.

Basically, the dream hypothesis is an attack on the reliability of our knowledge by suggesting that all we know might only be part of a dream. The literature gives this form:

1. Sometimes we take objects to exist in reality when in actual fact they exist only in our dreams.

2. No intrinsic difference exists between illusionary, dream experience and veridical, waking experience; therefore

3. We may be dreaming at any time.

Perplexity over the possible confusion between the real and dream worlds has not been limited to the Western philosophical tradition. The Taoist Chuang Tzu (third century B.C.) put the dilemma in this form: A man once dreamt that he was a butterfly fluttering among the flowers of the spring forest. Suddenly he awoke. Was he a man who had dreamt that he was a butterfly or was he a butterfly who was then dreaming that he was a man? The dream hypothesis creates two states: one the world of reality, the second a dream world in which all of the available facts are illusions or creations of the mind. Presumably we are in one of these two states at a given time but have no basis to tell which. This brings into question the reliability of knowledge, the reality of the world itself, and our perception of it.

Argument of the Génie Malin

With the suggestion of a malicious demon, or *génie malin*, Descartes (some would credit Montaigne with this invention) raised skepticism to heights undreamed of before. Since we will need to consider the specific characteristics of this demon when a distinction is made between a heuristic and skeptical view of the world, only a bare indication of his powers will be given now. Descartes granted his creation truly extraordinary abilities. This demon was dedicated to and capable of deceiving the human in all matters. He was able to distort man's information and faculties for judging it in such a way that what seemed the most certain, the most self-evident, and the freest from doubt was, in fact, false and untrue. Whereas illusions falsify the information of the senses, whereas dreams falsify the knowledge and reality of the world, the *génie malin* falsifies man's ability to judge. Many, perhaps most, persons believe in the existence of a benevolent God. How can we ever be sure, on the other hand, that Descartes' grand deceiver does not in fact also exist? We can never be sure; we can only hope that he does not.

Argument from Hypnosis

A particularly thought-provoking form of hallucination is the one induced by hypnosis. Purely philosophical arguments are not persuasive for many engineers. In my classroom I knew I needed to create an immediate, personal exercise to compel agreement with my central claim among those for whom the physical world has a particularly strong hold. My fatal blow to certain knowledge based on perception is hypnotism. I am convinced that hypnotism will fell the most recalcitrant engineer among us. Most of us were first introduced to this form of hallucination by the stage hypnotist whose act consisted of making members of the audience quack like ducks, sing Elvis songs, think they were naked, know they were the drum major of a band of seventy-six trombones, and do other inane and embarrassing things. As a result, we have come to discount hypnotism as a gimmick, a fraud, or a curiosity. This is unfortunate because hypnosis in actually used extensively by hypnotherapists to eliminate pain, help students study more efficiently, and eliminate harmful habits. As we will come to see, hypnotism has profound implications for the certainty of our senses.

Under hypnosis a person may be induced to see an object where one is not present or to deny the existence of an object where one is undeniably present. The subject is in a state of focused awareness or heightened suggestibility and is prepared to respond to the demands of the hypnotist. That this effect may be made to persist in the waking state in what is called a posthypnotic suggestion is less well known. Like the previous case of hallucinations, the illusions caused by hypnosis are completely integrated into veridical experience and indistinguishable from present reality. In fact, the subject may be led to believe and to staunchly defend situations that violate known laws of nature. To complicate the matter, a person in the waking state can be made to forget that he was ever hypnotized.

Two hypothetical examples of hypnosis are germane to this discussion. If a hypnotized person is told that he sees a book in his hands, experiment shows that he will report that he sees one. In fact, if he is told the preposterous fact that he holds the very first Gutenberg Bible ever printed, he will rationalize the existence of the book, plead his possession of it in the most convincing matter, and take exaggerated care in its handling. How can you be sure that you were not once under hypnosis, made to think that you are now holding a book, and made to forget that you were ever hypnotized? So much for your absolute conviction that you are holding a book at this moment to which I eluded when we began.

On the other hand, if a person under hypnosis is told that he can no longer see his hand, experiment shows that in fact he will no longer be able to see it. So much for Dr. Moore's hand referred to earlier when we considered the nature of the certain. As you recall, Dr. Moore raised his hands and said, "Here is one hand, and here is another." His claim was that any argument proposed to refute this claim was more suspect than the propositions themselves. I am not limited to abstract propositions to refute Dr. Moore; hypnotism provides me with experimental evidence that a hand may be seen where one is not present. Once again, doubt has retaken a beachhead once thought secured by the certain.

Over many years I have invited clinical psychologists, some of them from the Student Health Center of The University of Texas,* into my classroom to make students question the certainty of their senses. Initially, the experiment was double blind in that I had never met either the hypnotist or the student who served as the subject. In one particularly memorable year, I videotaped the class period in which a psychology major named Dixie was given a posthypnotic suggestion that made her *see* her dog named Cappuccino when the psychologist knocked twice on the table. In addition, the dog was to disappear when a book was dropped where the dog was located. Before twenty of her fellow students, when the prearranged cue was given, Dixie did in fact see Cappuccino. Since this was a posthypnotic suggestion and she was awake, the students could ask her questions. I remember their asking if the dog was solid and had a back side. She got up, checked, and answered "yes." They asked if it was wiggling. Her answer was "no." Repeatedly throughout the class period the hypnotist made the dog disappear and reappear. In desperation, one of the engineering students asked her how she could claim to see a dog when she knew in fact that it was not there. I could not have wished more of her. Her answer: "I really see that dog." Most students who attended the demonstration, who met the hypnotist and the subject, and who talked with the subject under posthypnotic suggestion came to doubt their senses.

I had occasion to give a lecture on the generalization of engineering method to universal method to an audience consisting of three or four hundred engineering professors.† The local university arranged for one of their hypnotherapists‡ to aid me in my presentation. Once again, I had not previously met either the hypnotist or the subject of the experiment. This time the student was induced to *feel* a glove getting hotter and hotter as she wore it when, in fact, it was a normal glove. Like the sense of sight, the sense of touch can be controlled by hypnosis.

In view of the examples of Cappuccino and the glove, what are we to make of the engineer's earlier claim to have designed a bridge and to be absolutely certain it exists? Recall his words. "I *see* my bridge, I *feel* my bridge, my bridge exists."

The argument for the unreliability of the senses based on hypnosis is not limited to the simple fact that humans may see an object when none is present. It is far subtler than that. The very existence of hypnosis opens the possibility that

---

*Drs. Don Reynolds and Ted Hill.
†Arlington, Texas, in 1986.
‡Dr. S. Fagg.

the human species may have been conditioned by a cruel environment in a past period of heightened suggestibility to see and accept a present world that does not in reality exist. Perhaps, hypnosis also instated the notions of arithmetic, mathematics, deduction, . . . , and other notions we hold dear. But, far more disturbingly, it may well have removed other more important concepts without leaving a trace. Until someone can—well—prove that this could not have happened, I am afraid that I must insist that perception and what we perceive are only heuristics.

I agree, however, that it is hard to accept the profound implications of these examples. One year a student named Alan found hypnotism too much to accept. As my class ended, he left announcing that hypnotism was a fake, that I was a charlatan, and that everything was not a heuristic. My answer was to remind him that if the demonstration of hypnotism were a hoax, it was a very elaborate hoax. If he denied the reports of a professional who made his living treating students at a major university, what warrant did he have to believe anything? If hypnosis did not exist, how could he be sure that Einstein, Hume, and all of the other famous people we had considered once existed? Unconvinced, Alan left and I did not hear from him for some time. One day about three years later, I heard a quiet knock on my door, ironically, as I was revising this book. Alan was at the door. He told me that he had been worrying about the hypnotist for years, that he felt compelled to drive across the state to tell me so, and that he had at last come to agree with me: *All is heuristic.* He then apologized, turned on his heels, and left. If you do not believe that hypnotism exists and are unwilling to face its implications, I have three years to wait for your knock at the door.

The conflicting opinion of philosophers, the tropes of Aenesidemus, synesthesia, the illusions, the dream hypothesis, Descartes' evil demon, and, finally, hypnosis—all raise unanswerable questions and require that perception be considered as a heuristic. To the extent that your belief in a perceived external world depends upon this heuristic for support, it is firmly established as a phantom, perhaps a phantom beyond exorcism, but a phantom nonetheless.

Until we can sort out these problems, I must now hold that any argument that depends on perception must stand as a heuristic argument to add to my growing list.

Experto Credite

We have now reviewed all but one of the woof heuristics promised. Most of them have depended heavily on the testimony of experts. This was necessary because none of us has the resources to become knowledgeable in all of the areas needed in this discussion. We had to "believe one who has had experience" in mathematics, philosophy, psychology, and physics because it was impractical for us to stop and become an expert in each of these areas for ourselves. Depending on the expert does have its drawbacks. I do not refer, as you may expect, to their fallibility. History clearly teaches that experts have often been proven wrong, and few are unaware of this fact. My concern is that experts suffer from the same lack of resources as do you and I when it comes to the next phase of our work, which is to generalize our results. We touched on this problem earlier when our loom was being strung with the promise to consider it in more detail later. Now the problem of relying on experts to make further progress becomes critical. Our problem is that experts, like us, do not have the time to become knowledgeable in all of the areas needed to justify the generality of my claim. They effectively become captured by their own studies and often confuse expertise in manipulating the heuristics in a narrow sota with expertise in manipulating those in the overall sota of humanity. This problem of the captured expert requires further elaboration.

Earlier, the intellectual countryside was divided into separate fields of study. Arithmetic, logic, truth, and so forth were removed, studied, and ultimately identified as heuristics. Now we must scan the whole landscape as from an airplane to recognize that everything is a heuristic, not just those areas we have chanced to investigate.

What we see as we look down are experts in each field prospecting for gold. Each prospector spends his life and resources digging deeper in his own plot of ground, for the most part wholly unacquainted with his neighbor who works the claim nearby. The mathematician studies mathematics; the physicist, physics; the philosopher, the theory of knowledge; and the theologian, religion. But in using all of his resources to distinguish himself as an expert in history, the historian must remain at the surface in logic, physics, and mathematics; the logician must remain superficial in history, theology, and physics; and the physicist must remain perfunctory in art, literature, and history. To become an expert, the vision of each must be limited to his own studies.

Occasionally a prospector will encounter doubts in his mine. The mathematician becomes aware of Gödel's Proof. The physicist, who owns the lease next door, studies quantum mechanics and accepts that theoretically he may disappear suddenly from where he sits to appear on the moon. The historian and cultural anthropologist examine the widely divergent opinions held in different ages and cultures and become suspicious of any claim to the absolute. The philosopher knows the problems of logic, of causality, and of truth; the theologian, the variety of religious experiences. Unable to appreciate the doubts in neighboring disciplines and often able to rationalize or minimize the doubts in his own discipline, the expert fails to generalize his doubts over all knowledge. Instead he generalizes his successes and falls into the egotism of the dogmatist. Soon the expert, captured by his own field, becomes convinced that his vision is an absolute characteristic of the whole countryside. Thus Hegel becomes a captured Hegelian logician; Kant, a captured Kantian philosopher; Skinner, a captured Skinnerian psychologist; and Freud, a captured Freudian psychoanalyst. Experts of limited vision are all too prone to mistake pyrite for true gold.

Since I am unprepared to accept the testimony of any scholar who seeks to refute my view that *All is heuristic* if he is not at least sota philosophy (Eastern as well as Western), sota science, sota theology (again, Eastern as well as Western), and sota technology, a very tall order to be sure, I need frequently the final warp

HEURISTIC: Beware the certainty of the captured expert.

The difficulty in finding an expert in "whether or not everything is a heuristic" is the major deterrent to depending on experts to establish that *All is heuristic*.

Argument as Argument

Even if an expert in "whether or not everything is a heuristic" were to exist, I cannot imagine how he would establish his claim. Please indulge me one last time to summarize the state of argument. With

1. All arguments based on induction as heuristic arguments (see page 65),
2. All arguments based on language as heuristic arguments (see page 122),
3. All arguments based on arithmetic as heuristic arguments (see page 130),
4. All arguments based on mathematics as heuristic arguments (see page 144),
5. All arguments based on deduction as heuristic arguments (see page 144),
6. All arguments based on logic as heuristic arguments (see page 152),
7. All arguments based on science as heuristic arguments (see page 173), and just now,
8. All arguments based on perception as heuristic arguments (see page 177),

am I not justified in writing argument itself as the

HEURISTIC: Argument?

All Is Heuristic

By now we have grown accustomed to using heuristics to explain and compel belief in the heuristic, so, once again, reusing one of the most ubiquitous heuristics *induction* to compel belief that *All is heuristic* should be no cause for alarm. Just as it becomes reasonable to guess that the sun will rise tomorrow after seeing many sunrises and that the next ostrich will have a long neck after many trips to the zoo, if enough concepts become tainted with doubt and are reduced to heuristics, eventually wagering that the next concept examined will also turn out to be a heuristic is the odds-on bet.

Some philosophers argue that the quality of an inductive argument is improved by judicious choice of example. They feel that the conclusion that all ostriches have long necks is improved if the ostriches are not chosen from the same area, but come from all regions of the world. I do not care to enter into this dispute because for me we would just be fighting over one more heuristic. I do point out, however, that the examples chosen to make the point that *All is heuristic* were not chosen at random but were selected as being what most people would agree are important, even fundamental, concepts. If all of these concepts are in doubt, is it not a good heuristic to doubt the less fundamental? Induction, being the heuristic that it is, does not prove *All is heuristic* no matter what the quality of the examples is, but somehow it does seem to prove it.

For induction to work its magic, a sufficient number of particular cases must be examined for a generalization to become compelling. From page one I began my collection of heuristics, surreptitiously; in the present section I have added to it, explicitly; and I could even now add further examples, indefinitely. At some point we must succumb to an inductive argument. When *classification* was first mentioned in the Introduction, the failures of the categories alive-dead, plant-animal, and male-female were cited. This casual reference was meant to evoke the enormous literature that includes Aristotle's ten categories, Kant's twelve, Hegel's three, and the indefinitely numerous ones of Ryle for the philosopher. Based on this volume of literature, *classification* was my first candidate for a heuristic. Soon an entire slate of potential heuristics was proposed, including *change, definition, set, ethics, induction, time,* and *language.* With each reference, what must have appeared as gratuitous pedantry to the casual reader was in reality busy subliminally planting the seeds of doubt. In the present section it is no longer a question of ruse. *Arithmetic, mathematics, deduction, certain, position, logic, truth, progress, causality, consciousness, physical reality, science, perception,* and, lately, *argument* are added to the list of heuristics. Although we have no more time for further discussion, sufficient evidence exists to question the concepts of *infinity, memory,* and the *existence* of other minds, but the list could go on and on.

For the engineer, I end with the fatal blow. He will no doubt be alarmed to learn that, in addition to all of the heuristics we have just examined, modern physics and statistical thermodynamics cast doubt about the extent to which even the laws of thermodynamics should be considered as "laws." Mass-energy is simply not always conserved, contrary to the insistence of the first law of thermodynamics. In certain nuclear reactions that are said to take place "off the energy shelf," over short periods of time the first law is not obeyed. Likewise, the second law of thermodynamics becomes suspect as an absolute because ensemble theory in statistical mechanics applied to an assumed random distribution in an adiabatic chamber predicts the possibility of a contradiction of it.*

There you have it. I cannot look at another ostrich. We have analyzed a large number of critical heuristics and synthesized an overreaching conclusion. I can do no more to accomplish the second major objective of Part IV, which was to make compelling the claim that *All is heuristic.* The particular warp and woof heuristics I have used were chosen somewhat arbitrarily, and others would have served as well. To think otherwise skirts perilously close to dogma. The intent has been to model the interaction and interdependency of all heuristics in a limited number of heuristics. In actual fact, the heuristics of the liberal arts weave imperceptibly into those of engineering that, in turn, weave into those of philosophy, into those of religion, into those of science, of history, of art. The basic strategy has been to bring into question concept after concept, all the while avoiding either an exhaustive treatment of too limited a number or a superficial analysis of too large a number. You are, of course, encouraged to choose others or to study those I have selected in more detail. Since your sota is fundamentally different from mine, it is understandable that different sets of heuristics might have been more compelling. For my part, the heuristics that have been examined convince me most profoundly, intellectually, emotionally, even dogmatically that everything is in doubt, that everything is provisional—that *All is heuristic.*

Befuddled?

By far the greatest tyranny over man is the idea that he can know anything. Certainly no necessary reason and necessarily no certain reason has been given to enable you to know that everything is heuristic. The nature of the situation is such that this is impossible. In the sweep of eternity, what are the odds that we presently know anything of cosmic import? Literature has always thought highly of the hero who struggles valiantly against his fate—the more inevitable

---

*Per Dr. Dendy Sloan.

the fate, the more noble the struggle. When at last the hero succumbs, we bestow upon him a certain grace. In the name of this grace, I suggest that you just capitulate, that you just accept, that you just refuse to struggle any longer and concede: I don't know; at best, I know. That is all we now know and given the present sota all we can conceive of ever knowing. Even if you must remain unconvinced by my tapestry and are either unwilling or unable to weave your own, will you not just capitulate, just accept, just know that the

HEURISTIC: All is heuristic

is, itself, a good heuristic? Curiously that is all I can ever ask.

Reduction of Kōan to a Preferred Form

The previous formulation of the central philosophical claim of this discussion is not entirely to my liking, and we must now become concerned about our commitment to its present form. *All is heuristic* depends too heavily on Western concepts, the syntax of English grammar, and contemporary rules of thought for my taste. Although we have used the peculiar convention of underscoring the second letter of crucial concepts to weaken any complaint, the present statement of the kōan still depends on one very special sota. Now we must clarify the constituent parts of the kōan to weaken further its hold on us. This will complete the second objective of Part IV of this discussion, which is to make the kōan compelling. What then, do we really mean by *All*, *is*, and *heuristic*?

All

Taking it seriously enough is the major problem in understanding the phrase *All is heuristic*. What, for you, may appear as a careless overstatement is, for me, an essential generalization. From *All engineering is heuristic* at the end of Part III, I pass to *All is heuristic* at the beginning of the present part of this discussion. Just as I refused to accept the first as an exaggeration, I am even less prone to accept the second as one. In both phrases, *all* means "all." If this one idea is thoroughly grasped and scrupulously observed, no problem will be encountered in the remainder of our time together. Because this extreme use of the word *all* is essential to our argument and so difficult to maintain for any length of time, let me emphasize—no, overemphasize—it by means of too many examples.

The set of all heuristics will contain all heuristics conceivable and inconceivable, spoken and unspoken, acceptable and unacceptable. Certainly all concepts, ideas, and words are meant to be considered as heuristics. This includes all of the concepts examined in the last section, of course. Likewise, *all* includes the traditional laws of thought as well as any and all distinctions, dichotomies, categories, and classifications. I also count as heuristics each inhabitant of that bizarre menagerie of creatures so dear to the heart of the philosopher: Buridan's ass, Quine's rabbit, Hempel's nonblack nonravens, demons (whether Maxwell's, Descartes', the brain in a vat, or those of my own making), and, of course, our old friend the camel's nose. Do not forget to include the philosopher's strange toolbox containing Occam's razor; Hume's fork, razor, and microscope; Moore's hand; the 99-foot man; and Goodman's oddly colored emeralds. You will have occasion soon to meet my own special guillotine to add to this list. I also take it as a heuristic. If *all* is really to mean "all," then we have no sound idea of logic, physics, substance, quality, action, passion, and existence—not as Francis Bacon has said because these happen to be ill defined at the present moment and are awaiting an improved method to correct this deficiency, but because they are heuristics and destined to remain so. As expected, we will find the heuristics of science—Newton's laws, quantum mechanics, and relativity—in the overall list of heuristics. Philosophy is also well represented through the views of the idealist, realist, nominalist, pragmatist, and skeptic, and so is religion, technology, and metaphysics.

What is less expected are other rather strange heuristics. For example, at one and the same time the overall sota that encompasses all heuristics contains these contradictory notions: There is an absolute and there is no absolute; This overall set is coherent (in the sense that all of the heuristics it contains depend in some way on each other) and the overall sota is incoherent; This set of heuristics is infinite and it is finite; and, finally, Everything obeys the law of contradiction because it does not obey this law. Indeed the overall sota enshrines the oxymoron. If we search hard enough, we will find the concepts: the nondifferentiating differentiation, icy heat, and the sound of one hand clapping. Granted I don't understand what these heuristics mean and perhaps neither do you, but the overall sota is not limited by our thoughts on the matter.

All includes the heuristics of all cultures, all times, and all places. Of course, the heuristics need not be expressed in the Western tradition nor represent exclusively the Western vision or Western language. The heuristics loosely called Chinese philosophy are all there, expressed both in the heuristics of the Chinese and in the heuristics that each alien culture uses to try to understand what Chinese philosophy is about. That is, the heuristics that comprise Eastern mysticism in a form acceptable to the various schools in the East are included, but also the gibberish Westerners use to try to understand the "Tao that cannot be expressed." The Zen master will be pleased to find that the heuristic "You should not be caught in any of the four propositions of *catushkotika**" is in the overall sota, but unless he is careful he will be displeased to find that the heuristic "You should try to become entrapped in the *catushkotika*" is also present.

We must not forget to include the concept of the heuristic, the sentence *All is heuristic*, and the notion of an overall list of heuristics in this overall sota. Although I hate to admit it, we must include the heuristics that there is no such thing as a heuristic, that the basic claim of this book is vacuous, and that an overall sota does not exist. Needless to say, you may consider some of the heuristics on this list better than others. In other words, given the heuristic *better* in your sota, you may think some of the rules of thumb in the overall sota are better, more efficient, more uplifting, or more appropriate. You may even feel that some heuristics are absolutely better than others or that some of them are silly, trivial, or frivolous. But, for inclusion in the overall list, all heuristics are equal qua heuristic. What can I say, what can I do to convince you of the extent to which I carry my claim? Never bother to ask, "Do you intend to include 'such and such a thing' as a heuristic?" I give you my answer in advance—it is *yes*. *Ad rem, all* means "all."

---

*1. It is A; 2. It is not A; 3. It is both A and not-A; and 4. It is neither A nor not-A.

The problem of taking the word *all* as seriously as I would have, and hence recognizing the ubiquity of the heuristic, is minor, however, compared to the difficulty of maintaining the position that everything is heuristic in the course of an argument. Just a moment's inattention is sufficient for a word, concept, or logical connective that is provisional, culture bound, temporal, and heuristic to be elevated to that of an absolute. (Even the word *absolute* in the last sentence suffered this fate, as it became—well, absolute.) If *all* is forgotten in the course of an argument, even for an instant, the camel's nose will get under the tent. Many an argument has faltered when the heuristic nature of a single concept has been forgotten, and a multitude of absolutes has quickly gained admission to the discussion. As we consider the self-referential nature of the kōan, the assumption that it implies excessive tolerance, and the implication of mysticism, I warn that the camel is looking for the tent. These examples will illustrate the orneriness of this beast.

The self-referential character of *All is heuristic* is immediately obvious. If the universal quantifier *all* is to apply to everything, then it must certainly apply to the basic claim itself. Mindful of problems of self-reference we saw with Gödel's Proof, even the freshman philosophy student is ready to pounce. Since the sentence *All is heuristic* must refer to itself, it is a heuristic and *therefore not universally true*. Unfortunately, the satisfaction our freshman feels is unwarranted and is due to his not being able to keep all concepts up in the air and heuristic at the same time. Like the juggler who holds on to one ball overly long, his act soon falls apart and absolutes rain down on his head. In this example, a virtual caravan of camels is begging for entrance. If *All is heuristic*, what are the concepts *self-referential, universal quantifier,* and *true,* not to mention the assumption of a specific system of logic that is needed to justify the passage to the phrase "*and therefore not universally true,*" but heuristics? I grant that it is difficult to ignore the problem of self-reference and hard to conceive of everything being a heuristic, but my present concern is not what the phrase *All is heuristic* means. In effect, I'm interested in the essence of juggling, not in convincing you either that a juggler exists or that it is desirable to invite one to our party.

A second example will show the persistence of the camel's nose. Some have argued that the phrase *All is heuristic* implies that all beliefs are relative, that nothing is certain, and that all is admissible. In related philosophical systems, Bayle has argued that radical skepticism and Locke that philosophic doubt would justify universal tolerance. Bayle, Locke, and my critics turn out to be rather intolerant themselves because their arguments require that the words *relative, certain, skepticism, doubt, justify, universal,* and *tolerance* be beyond doubt. I make the claim *All is heuristic,* but make no claim as to which heuristics you are compelled to believe, as to which heuristics must exist in your personal sota, as to which heuristics you must consider as good ones. While I might consider that "all values are relative," that "nothing is certain," and that "all is admissible" are good heuristics, attribution of these sentiments to me exclusively on the basis of my kōan is unjustified. If a person extends my claim in any way, that is his problem and is a reflection of his sota, not of mine.

As a final example, some have tried to equate the heuristic position with mysticism, a transcendent philosophy, or a philosophy of incomprehensibility. Those who would do so are warned that at least the nose, and quite probably the whole head, of the camel is already under the tent. Each of these notions is only a heuristic—and in my sota not a very good one at that. To understand the phrase

All is heuristic we must accept and keep constantly before us the absolutely essential point that in it *all* means "all." These three examples show how difficult this is to do in the course of an argument. Scrupulously observed, *All is heuristic*, where *all* means "all," inoculates us against the implacable dogmatist.

Is

Not much needs to be said about the second word *is* because its linguistic status is generally conceded to be ambiguous. In English, not only does *is* affirm and deny, but it can also ascribe a property, a relation, class membership, and, most controversially, existence or being. Philosophers have long sought its true function and arrived at little consensus. They ask such questions as:

1. "How can the difference between ascribing existence to a subject and ascribing a property to a subject be characterized?"
2. "Is 'is' ever a predicate?"
3. "If so, what sort of predicate?"
4. "In the simplest form as a statement of identity, that is as A is A, does the 'is' express equality or is it simply a copula or connecting link?"
5. "If a copula, does it imply the existence of the subject?"
6. "What does it mean to say of something that it is?"

Given the ongoing controversy over the word *is*, at the very least, we must hold that *is is is*.

In fact, some languages get along quite nicely without the word *is* and frankly so can I. The intent of *All is heuristic* is to bring the emotional feeling of the heuristic *all* into conjunction with the emotional feeling of the heuristic *heuristic* using much the same strategy as the Chinese language. The form *All; heuristic* is, therefore, a much improved approximation to my meaning because it recognizes the ambiguity in the parts of speech *all* and *heuristic* and does not prejudice the connection between them. Written in this way, the basic claim serves as a constant reminder that the grammar of the heuristic is not necessarily conventional. But we must retain the expanded form, at least in conversations with those who have not accompanied us this far, because it represents only a small change in the sota of standard English and therefore is more easily accepted and remembered.

Heuristic

In Parts II and III, the heuristic was explained in a variety of ways by comparing it to the instructions given to a novice chess player, by trying feebly to define it, by giving the signatures by which it may be recognized, by listing its synonyms, and finally by sampling typical engineering heuristics. Although at the time I recognized that definition was a heuristic, I defined; although I was aware that example was a heuristic, I gave examples; although I was certain that rational thought was a heuristic, I tried to be rational. At no time did I admit my subterfuge. Replacing define with define, example with example, and rational with rational, although dishonest, was acceptable because we were only concerned with a small subset of all heuristics, the engineering heuristics, and language and words could be used in their conventional roles. Questions concerning the status of words like *definition, plausible,* and *reason* lay outside of the sota of discussion and were never raised except by innuendo. Words could be trusted as uncontroversial starting points upon which to build, and the various strategies of explanation gave us a good first cut at understanding the meaning of the word *heuristic*.

Things have changed. Now as we try to improve on the first cut at the definition of *heuristic* and to understand this third word in the fundamental kōan, we find that the sota of discussion has expanded to include all heuristics and that words have lost their innocence. We must frankly admit that we are trying to say the unsayable. As a result of this new situation, the earlier efforts to explain the heuristic must be revised. Our chosen strategy will first examine the grammatical similarity in the predicates *being* and *heuristic* and then use words to suggest how you might manipulate your sota to come to feel the essence of the *heuristic*.

Linguistic orthodoxy holds that all words with the exception of *being* tend to classify things. They divide the world into objects of the sort named and everything else. *Chair* sets up a class of objects and divides the world into things that are chairs and things that are not. In other words, names apply to all things of a certain type, but not to all things of any type. In conventional practice, only *being* is common to all things of every possible kind. Take an object and remove from it all characteristics that distinguish it from at least one other object. It is now left with its being—informally, the fact that it *is* or that it exists in some sense. When this procedure is applied to *being* itself, to its negative *nonbeing,* and even to *nothing,* each is left with *being.* Among specialists, *being* is unique in this way, so unique that some question whether it should be called a predicate at all.

Contrary to conventional wisdom, I claim that not only the predicate *being* but also the predicate *heuristic* have unrestricted universality and apply to all things of every kind. If all characteristics of an object that distinguish it from at least one other object are removed, it is left with its heuristic (nature)—informally, the fact that it is tentative, provisional, based on a human sota, and plausible. To complete the parallel with *being,* the *heuristic,* its negative *nonheuristic,* and even *nothing* are heuristics.

Immediately, the logician will ask us to identify the characteristic that distinguishes *being* from the *heuristic* "when all characteristics 'that distinguish an object from at least one other object' have been removed," and then conclude that either the two are equivalent or that one or the other does not exist, with logic unable to choose which. In his haste, the logician overlooks the camel's nose. Although *being* would most certainly grant *being* to the concept *heuristic,* and the *heuristic,* to return the favor, would claim that *being* is a *heuristic,* the two positions are not truly symmetrical. The *heuristic* has an additional peculiarity and power that sets it apart from *being.* Aside from the uninteresting possibility that the two concepts are the same, a person who believes that *being* remains when all else is removed does not question the absoluteness of his claim, his method of stating it, or the logic system he uses to arrive at this conclusion. On the other hand, the person who endorses the *heuristic* as the remaining characteristic when all else is removed is more tolerant. If the heuristic is a characteristic of everything, then it is also a characteristic of his claim, his method of stating it, and the logic system he uses. Specifically, he recognizes as a heuristic "the characteristic that distinguishes *being* from the *heuristic* when all characteristics 'that distinguish an object from at least one other object' have been removed." I might add that, even more significantly, he also recognizes as a heuristic any efforts to make something of this supposed inconsistency.

The similarity between *being* and *heuristic* does not extend beyond their peculiar nature as predicates, and this comparison is more tactical than fundamental. The *heuristic*, like *being*, is peculiar because of its type of predication, but it is even more peculiar because like a hungry ouroboros,* it gnaws away at its own underlying substance.

This comparison of *being* and *heuristic* suggests the only strategy I have been able to devise to give a feeling for the *heuristic*. It works by successive approximations and uses words only as suggestions as to how you might manipulate your own sota.

For the minute, I do not ask you to believe that anything is a heuristic, but only to pretend that you do. Pick something easy to examine, say, the customs of a foreign country. It should not be too difficult to consider them as provisional, temporary, nonabsolute guides for organizing a country. Let us use the word *heuristic* for this description. While we are about it, many, if not all, of our native customs should also be taken as heuristics. Now consider the things you see, feel, and sense and imagine that they are uncertain. That is, assume that your senses are fallible. Make provisional the concepts of time, space, causality, and knowledge. That's a bit more difficult, but then we are only imagining. Now focus on language, particularly the words provisional, temporary, and nonabsolute. Are you sure that consciousness and the external world exist? Consign them to the fire for masquerading as absolute truth. Oh yes, the words *absolute* and *truth* in the last sentence must also burn. Avoid carefully the trap of mysticism, the Tao, and the undifferentiated continuum. We must pretend that they are also heuristics. Can you be absolutely certain that other people exist? Specifically, what do you really know about me? Your knowledge of a concrete author sitting opposite you in this discussion is only based on heuristics. Let's assume that I am just one more sota. Consider *being*, even your own being, as only the most persistent dromedary. Continue this procedure until everything, that is *all*, is imagined as a heuristic—and then allow the very notion of the heuristic to fade away ever so slowly like the grin of the Cheshire cat. . . .

With this final act of will, the second of the seven objectives set for Part IV is completed. It is subtle enough that now would be a good time to follow the advice given in the Introduction to this discussion: put this book down—and rock.

After admitting the inherent difficulty in trying to explain the heuristic with heuristics, we chose one small subset of all heuristics, the English language, and examined each of the components of the phrase *All is heuristic*. What we found was that *all* means "all," that *is* may be safely ignored (although in informal conversations, it is probably best retained), and that *heuristic* is best understood as a characteristic of everything, including ultimately itself.

---

*An alchemists' symbol, a serpent with its tail in its mouth.

COMPARISON OF HEURISTIC AND SKEPTICAL POSITIONS

The preceding section sought to make a clear distinction between being and the heuristic. This section seeks to make an equally clear distinction between the heuristic and the skeptical positions in their relationships to knowledge.

I make no claim to having been trained as a philosopher; I am educated as an engineer. I do not pretend to be able to knit together complex arguments in the theory of skepticism. Therefore I will have to leave such erudite arguments as the proper translation of *aporia* and whether or not it leads to a state of frustration or detachment to my learned friends across the campus. Skepticism is so essential to my cause, however, that I must do my best.

A skeptic is a person who doubts and acts on that belief. The English language does not provide an equivalent word for a person who believes that everything is heuristic and acts on that belief. Therefore, I will continue to use engineering as a model and choose the engineer as a stand-in for such a person. All practicing engineers may not endorse some of the characteristics that will be attributed to the "engineer" in this context. I am not trying to redefine engineering, but to provide a convenient term to juxtapose with the term *skeptic* as we try to distinguish between the two.

Until now I have been hiding behind the skirts of the skeptic giving only an occasional glimpse of the true heuristic position. Because the skeptical and heuristic views are similar, because the skeptical arguments of the past have won all confrontations with the dogmatists, and because the acceptability of skepticism adds credibility to the heuristic position through kinship, as a first approximation, I have been willing to bask in the reflected infamy of the skeptic. Since we will soon discover that the skeptic and the engineer fare differently in critical, life-threatening situations, it is now advisable to come out into the open and make the necessary distinctions between the two. This is the third objective that was promised for Part IV.

There can be no doubt that the skeptic and the engineer are different. According to Pyrrho, the goal of the skeptic is peace of mind, unperturbedness, calmness untroubled by mental or emotional disquiet, or, in a Greek word, *ataraxia*. Things have not worked out the way Pyrrho must have intended. Instead of peace, the skeptic has caused a dread to descend on the land that is totally alien to the spirit of the perennially optimistic engineer. In addition to rejecting the final state that the skeptic has provided, the engineer also rejects his strategy for getting there. The decisive technique for achieving the Pyrrhonian peace of mind was suspension of judgment, or, to use yet another catchword of the skeptic, *epochē*. In engineering, suspension of judgment is nearly always a bad heuristic. To appreciate how the skeptic has gotten into his present predicament, we will look at a brief history of skeptical thought and then conclude with five specific examples that illustrate the differences in the skeptical and heuristic positions. This knowledge will allow the engineer to escape the skeptic's fate and add precision to the central claim of this discussion: *All; heuristic.*

History of Skepticism

Most of our knowledge of early skeptical tradition is from Sextus Empiricus, who distinguished between two varieties of skepticism. The first kind was Academic skepticism, which held that nothing could be known with certainty. The second kind was Pyrrhonian skepticism, which held the even more radical position that there was insufficient evidence to know even that much. The Pyrrhonians felt that the Academics were being too dogmatic when they affirmed the impossibility of certain knowledge and suggested that it was better to suspend judgment on all questions concerning knowledge, even the question as to whether or not knowledge is possible.

From this beginning, the dogmatist has been a moveable quintain but a skeptic has always arisen to do battle in defense of doubt's honor. For example, during the medieval period, when Father Garasse charged that Catholic Pyrrhonism was an atheistic plot, Saint Augustine concluded that skepticism could only be countered by divine revelation. Early in the Renaissance, Erasmus, a Christian skeptic, arose against Luther to defend the position of the Catholic Church by arguing for a suspension of judgment and acceptance of the church's view on the basis of faith and tradition. Somewhat later, Sanches and Montaigne emerged, the former to use classical arguments to show that scientific knowledge cannot be attained and the latter to provide a general skeptical critique of the possibility of human beings understanding anything. Montaigne concluded that the best policy was to follow the Pyrrhonian suspension of judgment, live according to nature and custom, and accept what God chooses to reveal to us.

The battle continued into the seventeenth century. When Charron, Gassendi, Mersenne, and others moved to reassert the skeptical conclusions, counterefforts began to establish a firm basis for knowledge. Most notably, Francis Bacon sought to replace the old Aristotelian methods with new methods so as to establish firm and unquestionable results, and Descartes sought to overcome all doubt and to find at least one absolutely certain basis for knowledge. Unfortunately for Descartes' program, his first published work, *The Discourse on Method*, was destined to be the first of a series of books that were to be considered more skeptical than any to date.

The next period we should consider in this brief history of skeptical thought is that of the so-called modern skeptics: Bayle, Hume, and Kant. Bayle has been called the most incisive of this group. His writings were dialectical stilettos to the corpus of all dogmatic argumentation, be it the knowledge claims of theology, metaphysics, mathematics, or science. Bayle professed to be against everything that is said and everything that is done. His thrusts proved far from fatal, however, and Newtonian science breathed new life into the dogmatist. While it was freely admitted that the scientist did not know everything at the moment, all the scientist asked was just a bit more time and it would be. To joust with this new adversary, a champion for the skeptic's cause arose in the person of Hume. Previous sections have emphasized Hume's influence in destroying causality, but he also presided over the weakening of man's supposed knowledge of substance, the self, and external existence and showed that no truths about matters of facts could be established deductively or inductively. Awakened from his slumber to the problem of knowledge by the work of Hume, Kant decided to take up the relay. His strategy was ingenious: He simply assumed that knowledge was possible, hence skepticism was false, and sought the conditions under which this might be possible. We first introduced Kant's system at the end of Part III when we were concerned about representing knowledge in a coordinate system. Soon we will revisit it once again to insist on its relationship to heuristics. For now it is sufficient to remember that trying to set up a system that guarantees certain knowledge has always proved an unprofitable business and that Kant's efforts proved no exception. Many now consider that Kant provided a new type of Pyrrhonism so radical that he has been called the greatest of modern skeptics.

Let us complete this trip through history by considering some recent skeptics. Work in the social sciences, notably that of Nietzsche, James, and Freud, shows that the views we hold are fundamentally influenced by what I would call economic, social, and psychological heuristics. That is, truth is relative to the sotas of a person and his environment. Mauthner is another more recent philosopher who arrived at a complete skepticism about the possibility of true knowledge from his work on language. And finally, the animal faith of Santayana, the skepticism of David Levi, and the existentialism of Camus must be mentioned in passing. Borrowing from Greek mythology, Camus identified the plight of modern man with Sisyphus, who was doomed to spend eternity pushing a huge rock to the top of a hill only to have it roll once more to the bottom. In Albert Camus' words from *The Myth of Sisyphus* "If this myth is tragic, that is because its hero is conscious. . . . [He] knows the whole extent of his wretched condition."

From Pyrrho to Camus, a journey of some twenty-three centuries—all to what avail? Nothing is too absurd, it seems. Do you want everyone turning into rhinoceros and stampeding through the living room? Just call on Ionesco: "It's not a certain society that seems ridiculous to me, it's mankind."

At the present, the analysis of the possibility of certain knowledge is quiescent without quietus. An uneasy truce has settled over the land and the combatants rest in despair, confronted by an absurd, meaningless, unintelligible, senseless, and ultimately Godless world. If this is the *ataraxia* promised by the skeptic, no respectable engineer would want any part of it.

As mentioned earlier and often noted by the authors who have just been visited, the decisive strategy for achieving this dubious peace is to suspend judgment. Since, according to the skeptic, equally good arguments can always be presented on both sides of any question, refusing to judge is considered the best policy. This policy has led many to conclude that excessive skepticism was not so irrefutable as impractical, for if adopted as a way of life, "all discourse, all action would immediately cease, and men would remain in total lethargy until the necessities of nature, unsatisfied, put an end to their miserable existence." One ancient skeptic, Cratylus, an extreme example to be sure, was reduced to just wiggling a finger because he was afraid of speaking an untruth. The engineer wants even less a part of the inaction the skeptic proposes than he does of the dread it produces.

Differences between Skeptic and Engineer

With a difference between the skeptic and the engineer assured by this brief history of skeptical thought, five examples begin our search for its nature. They include the tendency of the skeptic (unlike the engineer):

1. To ignore the limitations of his sota,
2. To cohere with the sota of the external world,
3. To give the home field advantage to the dogmatist,

4. To fall prey to his own arguments, and

5. To reify doubt.

The difference we seek lies in the common thread that runs through this collection of examples.

Génie Malin

Descartes' demon, or *génie malin*, is a good concept with which to start as we try to understand how the skeptic ignores the limitations imposed by his sota. We have already made the acquaintance of this demon in an earlier section, but now we seek to know him more personally. He is considered by many to raise the final level of doubt and to lead to an ultraskepticism. Descartes describes his creation as follows:

> I shall then suppose, not that God who is supremely good and the fountain of truth, but some evil genius not less powerful than deceitful, has employed his whole energies in deceiving me; I shall consider that the heavens, the earth, colors, figures, sound, and all other external things are nought but the illusions and dreams of which this genius has availed himself in order to lay traps for my credulity; I shall consider myself as having no hands, no eyes, no flesh, no blood, nor any senses, yet falsely believing myself to possess all these things

What is intriguing about this description is that it is strangely anthropomorphic; the Cartesian demon is almost human. He is deceitful and evil, but we all know people who have had these traits—the *génie malin* is just a bit more duplicitous. He raises doubt, but I know all about doubt. And almost kindly he limits his deception to matters I can understand such as the absence of colors, hands, eyes, heavens, and earth. What do I have to fear from this demon? I feel I know him personally. In ancient time the *génie malin* no doubt deceived people that the world was not really flat, that the world was not really made of water, that Zeus was not really the king of the Gods, and that there was really such a thing as rational thought. Heaven only knows what deception he has planned for the future, but I'm sure the people living then will understand what he is about. Although I do not know how he came to have his powers, he is hauntingly familiar to me and strangely tailored to the sota with which he must deal. Descartes' creation is a very mundane, a very state-of-the-art demon. I don't for one minute respect his powers.

My fear is of an even stronger, much subtler demon. Although somewhat similar to the Cartesian demon that is malicious and deceives, my demon is malicious and deceives. But the heuristic demon does much more: He _____, _____, and _____. Not constrained by any present or future sota, he is beyond existence and non-existence, good and evil, freedom and dignity, knowledge and skepticism, rational thought and irrational thought, doubt and certainty—beyond even the heuristic and the non-heuristic. All the *génie malin* and I have at our disposal is a rather motley collection of heuristics, rules of thumb, hints, hypotheses, axioms, and principles—ah, but this new demon, he has _____ at his. When my sota does not even allow me to conceive of the demon's powers, when I cannot even understand who he is and what he is doing, now there's a real cause for fear. The difference in the Cartesian demon and the heuristic one reflects the difference between the skeptic and the engineer. While the descriptions of both demons are necessarily limited and constrained by the current sota, to describe this new demon we must recognize this limitation, acknowledge its power, and imply an extrapolation beyond it.

Coherence

Unlike the engineer, the skeptic is not sufficiently affected by his doubt of doubt. This is the second striking difference in the engineer and the skeptic. Since the skeptic is not surprised that his actions always remain coherent with the sota around him, his behavior gives him away. If an avowed skeptic is asked, "Do you doubt everything?" without hesitation he will answer "yes." If he is asked, "Does a skeptic suspend judgment on all questions?" equally quickly he will answer "yes." Finally, if he is asked, "Will a true skeptic assert anything?", again with no pause, he will answer "no." The skeptic seldom, if ever, equivocates in answering. If he truly doubted the concept of doubt, we might at least expect him to stutter a bit when asked these questions as he vacillated between all of the possible permutations of the object in doubt and whether or not to doubt it. We would expect an occasional unexpected combination such as, "I never doubt; I'm a skeptic." A skeptic's answers always cohere with the expectations of his listeners, that is, with the expectations of the sota that confronts him. A theory based on extreme, complete doubt would admit this observed consistency only as a very unlikely act of chance. The skeptic may march to a different drummer, but he somehow manages to remain in step with the old cadence.

On the other hand, a theory based on the heuristic expects this coherence because although coherence and incoherence are both heuristics, in today's world, coherence is usually considered a better one. While, of necessity, both the skeptic's and the engineer's sotas cohere with the world around them, the skeptic should worry about this fact. Because the traditional skeptic does not occasionally *exhibit* his doubt of doubt through his actions, he appears more dogmatic than skeptic.

This is not to argue that the skeptic has never doubted doubt—the Pyrrhonians, in particular, distinguished themselves on that point. Nor is it to criticize or try to refute skepticism by claiming that it is internally inconsistent, as did Sorel, Bolzano, Bayle, Hume, and others when they noted that the skeptic who claims to know nothing is contradicting himself in claiming to know that. The inherent difficulties of stating the skeptic's position without self-contradiction have long been recognized, and Montaigne and Santayana have definitively answered these charges. The first reminded us that the Pyrrhonians needed a nonassertive language in which to state their case, since as soon as they asserted anything, opponents could claim that they had violated their own view. The second reminded us that the skeptic is not committed to the implications of other men's languages; nor can he be convicted out of his own mouth by the names he is obliged to bestow on the details of his momentary vision. Indeed, my acceptance of these concerns motivated my decision to underscore the second letter of words that might deceive us in this way. My problem is not with a lack of a conviction on the part of the skeptic or even with a lack of a mechanism to express this conviction, but with a lack of chance incoherence when he is so constrained. Even though the skeptic is shackled to the sota that defined his opponent's language, his struggle against his bonds leaves no marks.

Home Field Advantage

The third difference between the skeptic and the engineer is that skepticism feels obligated to give the home field advantage to the enemy to be accepted as a philosophical system. Unfortunately, the skeptic has always won under these adverse conditions, and although he may fuss a bit, he does not refuse to accept them. A philosopher is not interested in a position held capriciously, but only in one supported by argument and reason. The skeptic is made to have reasons for his doubt and to limit himself to the game of the rationalist in order to be invited to play. In so doing he is made to beg the major question in dispute, the existence of valid grounds for knowledge of any kind. Commenting on this peculiar situation, Hume writes, "It may seem a very extravagant attempt of the skeptics to destroy *reason* by argument and ratiocination, yet this is the grand scope of all their inquires and disputes." In effect the skeptic invites the camel's nose (in the form of admitting the existence of rational argument to obtain knowledge) under the tent and then complains too loudly that it is there. As we have seen, the engineer who accepts that everything is heuristic accepts no such limitations. He is willing to make his points by cajoling, coaxing, and wheedling, at best a counterfeit form of argument.

Skeptic's Pride

The distinctions between skeptic and engineer do not end here. The skeptic is far too smug about his victories. Once the skeptic has consented to argue, he does not remain skeptical about the force of his own arguments and is taken in by them himself. For example, the skeptic takes obvious pride in his *diallelos*. Recall that the *diallelos* was that formidable skeptical weapon that purports to destroy the notions of truth, proof, and so forth by insisting that each must depend on an unproven assumption, an infinite regress, or a circular argument. As you recall we have already welded this weapon against the notion that language can explain everything. What right does the skeptic have to find his victory convincing? I cannot think of a fourth alternative to those proposed in the *diallelos*, but that is my weakness, because of the limitations of my sota, and is open to renegotiations in the future. As you recall, we avoided this problem in our own argument by insisting that we weaken it and write *diallelos*. Unfortunately no skeptic writing on the matter feels obligated to do the equivalent. He always seems to be taken in by his own arguments. The same lapse is seen in any satisfaction the skeptic derives from his tropes, *epochē*, or, for that matter, just sitting quietly and wiggling a finger. The engineer both accepts and rejects both the heuristic of doubt and the heuristic of certainty. He is prepared to act or not to act on either heuristic depending on the dictates of the moment. If you became overly concerned and doubted arithmetic when you read the discussion of Gödel's Proof, you are a skeptic; if you didn't get overly exercised one way or the other and considered it just one additional, good heuristic, you are an engineer. For a skeptic, it is with his pride, not after it, that his fall goeth.

Reification of Doubt

The doubter's dependency on doubt for his existence is a final distinction between the skeptic and the engineer. This reification of doubt is the private affair of each skeptic. If doubt, its synonyms, descriptions, and definitions were all removed from the English language, the skeptic might still feel the twinge he used to call doubt but he could not show us his skepticism by living in such a way that an outside observer could confidently say that he was being skeptical. Think what is meant by the claim, "I doubt." The skeptic might hesitate, alternate between positions, even suspend judgment, but if the observer is not told that the inner state of the skeptic is one of doubt, the observer simply observes the skeptic's behavior and concludes that he is hesitating, alternating, and suspending. The reason for this behavior (that is, that he doubts) can never be guaranteed. Just imagine a second person who does not doubt, but would have us think that he does. Like the first, he can hesitate, alternate, and suspend. We cannot convincingly say we know for sure that another person doubts, that another person is a skeptic. Doubt is a person's own assessment of his internal state. It is a heuristic in his sota and only accessible to an outsider heuristically.

On the other hand, the engineer cannot but live his life in such a way that an outside observer could say he is using heuristics. To say that an engineer uses heuristics is our assessment of his actions. It is a heuristic in the outsider's sota, not necessarily in the engineer's. Indeed, the engineer might even deny that he used heuristics. This discussion is a good example. After observing the behavior of many engineers, I am prepared to assert that engineers use heuristics at all times. Other engineers may disagree, may have never heard of a heuristic, or may assert that engineers only use them occasionally. But I can still assert (based on my sota) that the engineer uses heuristics. Or, to clinch the point, consider the one-celled animal called an amoeba. Can it live a skeptical life? A heuristic one? The skeptical life depends on the opinion of the avowed skeptic; the heuristic one does not depend on an avowed engineer. Skepticism is simply not skeptical enough for my taste—it depends too much on the reification of doubt.

Because the skeptic chronically ignores the limitations of his sota, is not embarrassed about cohering with the sota of the external world, gives the home field advantage to the dogmatist, falls prey to his own arguments, and reifies doubt, he is significantly different from the engineer who simply uses the doubtful, the provisional, and the heuristic.

I do not, however, wish to be overly harsh with the skeptic, for our dispute is only a family quarrel. As an engineer I cannot help but admire a system that has always represented an appropriately small change in the state of the art with which it has had to deal. Now, however, we must build on the foundation the skeptics of the past have secured.

Throughout the ages, daedal philosophers have been hard at work erecting a labyrinth of theories to protect *certain knowledge* from attack. Meanwhile equally skillful skeptics have laid down a thread of doubt to ensure a safe return from the twists and turns of reason after doing battle with the philosopher's creation. Although the beast is now at bay and sits cowering in the corner, the skeptic, sensing some of his own nature in it, has no heart for the coup de grâce. It is, therefore, for a new generation of engineers who alone can destroy labyrinth, thread, and beast to create a new *ataraxia* and a new world that we can all desire. This time a world based not on doubt, not on certainty, not on *epochē*, but on the heuristic.

An Impregnable Defense

One small matter does need to be taken care of before we can proceed. I do no one a service by cajoling him into accepting that A̲l̲l̲; h̲e̲uristic only to open him to a fatal attack. What kind of ally is it who presides over the total destruction of a newly acquired world? Hostilities against our claim provoked by the well-meaning, the confused, and the threatened are inevitable, and we share the problem with the juggler that one miscue, one moment of inattention will cause our act to collapse. Wary of this possibility, I feel obligated to provide a simple, powerful, sure defense guaranteed to ward off any attack if faithfully followed. It is the three-step

HEURISTIC: (1) Let the opponent speak first; (2) say in a gentle voice, "That's an interesting heuristic"; and (3) observe a long silent pause.

In implementing this heuristic, the final pause is by far the hardest part. The temptation will be great to join the wise men who rush in where the engineer fears to tread and argue, explain, and try to convince. Time will come later for compelling belief after the first skirmish has been won. The persistent repetition of "That's an interesting heuristic" followed by the pause will reduce even the most scathing critic to speechlessness—a condition of eloquent agreement with our cause.

For this defense to work, we must be steadfast in our resolve. We must be willing to accept anything and everything as a heuristic, although we are not required to consider it a particularly good one. If someone responds to our claim with, "That's the craziest theory I've ever heard of," we will calmly respond, "That's an interesting heuristic." To, "You have contradicted yourself with a self-referential statement and, therefore, your position is untenable," again we must say, "That's an interesting heuristic," followed by the long pause. I know this defense to work. I was once at a conference* with a professor whose specialty was Kant. Long into the night we (should I say, he) argued. In exasperation, he left for bed with the parting comment, "How can we make progress if you refuse to engage?" "That's an interesting heuristic," I said. "Sleep well." The next morning at breakfast I was met with his greeting: "I give up. You win."

As effective as this defense is, many will not have the fortitude to use it, will fall by the wayside, and will be lost because it is not as easy to use in practice as it is in theory. For example, try this defense on the three statements:

1. Accepting all as heuristics means that you tolerate all points of view, even Hitler's.

2. God is dead.

3. Weak babies should be killed at birth and elders who have outlived their usefulness should also be killed.

---

*MITF Project, CACHE, New Seabury, Massachusetts, August 19–21, 1975.

Only when we can answer, "That's an interesting heuristic" to each of these followed by a very long pause does our defense become impregnable.

OVERALL sota

Now is a good time to pause once again and reaffirm the line of this difficult part of our discussion. We have made compelling the notion that everything is a heuristic and differentiated the heuristic position from the skeptical one. Immediately, and with little effort, all of the results from the first three parts defining the engineering method may be taken over intact into the present considerations. This includes the sota, the definition of best practice, and the Rules of Judgment and Implementation. We could also consider the evolution and transmission of our new, all-inclusive sota with interesting results. This expanded vista is not an insignificant achievement and the panorama is truly breathtaking. Rather than reconsider all of these concepts in detail now, focus will be on several consequences of our new view. A later section will make a few comments on several of these earlier considerations within this new context.

By now we have grown accustomed to the acronym *sota* that has proven so useful in avoiding the ponderous engineering term *state of the art*. We have also come to accept the word *all* in its most encompassing form in the ungrammatical construction *All; heuristic*. Combining these two established notions in a new term *overall sota* as we did when we considered the nature of the word *all* will keep them constantly before us. By overall sota we mean, of course, the set of all possible heuristics. After examining this new concept, the way will become easier and the pursuit of universal method may begin in earnest.

To begin the hunt, we consider the fourth objective of Part IV that is to consider the overall sota in some detail. It will be found to be far less radical than it may appear at first because we find that many near synonyms for it have appeared throughout history. Reviewing the most important of these will add credibility to our new notion and show how it differs from what has come before.

The second step is to partition the overall sota into various subsets or subsotas. At this point we will discover that each of these smaller sets can represent a specific concept by looking at several examples including the system of Kant. All subsotas do not have distinct boundaries, but are what we might call *fuzzy** subsets or *fuzzy subsotas*. Three fuzzy subsets of the overall sota, those used to define *science*, *common sense*, and *man*, are then considered as examples.

The final step of this section is to consider our own personal fuzzy subset of heuristics. This will allow us to consider the compelling nature of the subsotas we each use to structure our world, to demonstrate the incoherence of our personal sota, and to consider the engineer's peace of mind or *ataraxia*. A peculiar guillotine and a group of condemned philosophers will serve us well in this last effort. Since both the general concept of method and the particular concept of

---

*The word *fuzzy* is not a frivolous choice, but a technical term used in an important branch of modern mathematics called Fuzzy Set Theory.

universal method turn out to be specific, compelling fuzzy subsets, this section will establish a framework for keeping the promise made in the introduction to define universal method.

Synonyms

Philosophers have produced a multilingual lexicon of near synonyms for the overall sota. Each of these alternative forms turns out, however, either to be a subset of the definition given here, to take itself too seriously, or to require a human as a starting point. Let us review these near synonyms now in order to have a greater appreciation of the term *overall sota* and to see where each falls short.

The two most familiar synonyms for the overall sota are *zeitgeist*, often translated as the spirit of the age, and its near cousin *volkgeist*, or spirit of a nation. The *zeigeist* or *zeit der geist* refers to the characteristic intellectual, moral, and cultural state of an era taken as a whole. Clearly, the heuristics that make up the concepts *intellectual*, *moral*, and *cultural* are contained in the overall sota. What is less clear is the relationship between the individual and the *zeitgeist*. Herder portrays the *zeitgeist* as a mighty demon to whom we are all subordinated, and Hegel insists that philosophy is the *zeitgeist* or intellectual trends of an epoch apprehended in thoughts. In either case, man is excluded from the *zeitgeist*, and the spirit of the age would seem to be a proper subset of the overall sota as we have defined it.

This feeling of detachment is evident in other near synonyms. Mannheim's *situationsgebunden* (situation-bound) in which knowledge is tied to a set of sociohistorical circumstances that evolve in time as each age develops a different style of thought; Wittgenstein's *language-games, world-pictures,* and *totality of judgments* that appear often in his later writings as the context in which meaning resides; and, perhaps, even Heidegger's *zeitalter* as an epoch in time characterized by a given interpretation of truth are specific examples. Each is a subset of the set of all heuristics.

Closer to the notion of an overall sota is the Japanese (and Chinese) word *kyogai (ching-ai)* loosely translated as environment. To the Zen master a person's *kyogai* is the absolutely unique, inner structure or frame of consciousness from which all of his reactions come and wherein all outside stimulations are absorbed. Clearly, the *kyogai* is similar to the sota of an individual. The major differences between the *kyogai* and the overall sota are a strange feeling that a person's *kyogai* is a thing and that it exists in some sense. It also uses the words *absolutely, unique, inner,* and *consciousness* a bit too dogmatically for my taste. We have consistently insisted that everything, and that would include a person's sota, is a heuristic. The same must apply to all of the words just mentioned.

A different distinction separates the word *weltanschauung* (often translated as worldview) from the term *overall sota* as applied to an individual. While it is clear that the heuristics a person uses to define himself and those used to structure his world could be part of his *weltanschauung*, it is hard to see what would happen to this concept if there were no person to have this worldview. The overall sota contains heuristics independent of any specific intellectual parse called human.

Other anticipations, approximations, and subsets of the term *overall sota* could be given and would surely include Whitehead's *climate of opinion* and Locke's *well-endowed opinions in fashion*. The essential point is that, although each of the philosopher's terms reviewed adds dimension and color to the concept of an overall sota, each fails as a true synonym. Some overemphasize the division of the individual and his environment. Some take for granted the absolute existence of the concept in question. Others require an individual as a starting point. That is, they are all heuristics themselves, and none have the power of the ouroboros.

Partitioning Overall sota

Throughout our discussion the overall sota was implicitly partitioned into subsets as the need arose. The heuristics that defined *engineering* were considered as a unit and used to examine that concept, and the heuristics that define *language* were considered as a unit when it became appropriate to consider that concept. You will also recall that we considered the subsets that represented *music* and *logic*, in addition to *engineering* and *language*, in Figure 30. This procedure of partitioning the overall sota into subsets must now be made explicit. We must also examine whether or not the boundaries around different sets of heuristics are distinct and the same for all people.

Concepts as Subsets

Analogous to the division of sota$|_{\text{eng.prof;t}}$ into subsets that represented individual engineers, the overall sota can be partitioned into many, often overlapping, subsets. These subsets are the basis for defining cultures, professions, and, indeed, all concepts.

Specific subsotas, for example, contain the heuristics that make the French, French; the Japanese, Japanese; and the Americans, American. Although a large overlap is expected in the heuristics that define the various cultures, this overlap is not complete and critical differences do occur. The subset that defines the traditional Eskimo culture, for instance, contains the heuristic that old people should abandon camp and go out into the winter alone to die once they become too much of a burden on the group. In Sparta unwanted children were left to die of exposure, and certain tribes of American Indians held the newborn under cold waterfalls in elementary forms of eugenics. Happily, from my point of view, these heuristics do not exist in the sota that defines my culture. Less significant, but more conspicuous cultural differences occur in clothing. The heuristics that define appropriate dress, or lack of it, are different for Americans, Japanese, and some African tribes. Personally, I have never thought that a gown of azure with green underwear and a hat were flattering, all the less so when the hat is in the form of a turban and surmounted with a red cross, but for some this is a costume fit for a utopia. A culture is completely known once the set of heuristics that define it have been identified.

In a similar fashion, we could identify the set called science, which could then be further subdivided to identify those heuristics pertaining only to physics and then partitioned once again to leave only a sota containing the heuristics of quantum mechanics. The layperson may not even understand the heuristics contained in this last division and usually views it as a sideshow peopled by unreal objects with strange-sounding names such as wave functions, S-matricies, and Hamiltonians. It is a world in which the everyday heuristics of time and space are not necessarily even present.

Literature stabilizes specific subsotas. Heuristics define a Walter Mitty, a Pollyanna, and (if we have the sota of a specialist in world literature) an Akaky Akakievich. When an author lists six heuristics to make people like you, twelve heuristics to win people to your way of thinking, and nine heuristics to change people without giving offense or arousing resentment, he is defining his sota for winning friends and influencing people.

The relationship between sets of heuristics and a specific concept is not limited to simple examples such as these. On a more profound level, human thought is itself an example of the use of various subsotas. The world around us is extremely complex and without collections of heuristics, we would quickly be overwhelmed by the vast amount of data. What we do instead is simplify the world. We categorize, define, create concepts, and develop aphorisms so that we can handle the situation. Just as everything in engineering is a subsota of the sota of engineering, all concepts, including both method and universal method, are subsets of the overall sota.

Instead of describing the world in terms of heuristics and collections of heuristics, many philosophers have tried to establish an absolute structure for knowledge. Since Kant's system is by far the most elaborate, monumental, and celebrated attempt to do so, we should consider it now and insist that we are only dealing with yet another complex subsota of the overall set of heuristics. Kant's rather elaborate system was first introduced on page 105 near the end of Part III. Now might be a good time to review this compact paragraph because the complexity and obscurity of Kant's system make it hard to keep in mind.

When we first discussed Kant, our goal was modest. We only insisted that Kant, like Einstein, like the engineer, was setting up a coordinate system against which to understand his world. As you recall, at that time, we slyly weakened Kant's hold over us by introducing the modern mathematical notion of turtle graphics and using it to dispense with all external coordinate systems. We were asked to imagine that we were riding on the back of a turtle and whispering heuristics to it. Now we continue the process of weakening Kant's hold over us by insisting that Kant's worldview is but one more subsota of the set of all heuristics.

Like the definition of the French, like the definition of physics, like the definition of a Walter Mitty, Kant's system is also a sota. We have already found sufficient reason to identify space and time as heuristics, but what else could his concepts of quantity, quality, relation, and modality, not to mention the bewildering array into which each of these can be divided, be but heuristics? Shouldn't Kant write mind for mind, *a priori* for *a priori*, and things-in-themselves for things-in-themselves? I agree that we seem to structure our experience in terms of something (Kant may call it a Table of Categories; I obviously prefer to call it a sota or set of heuristics), and I might be persuaded that the set he proposes is a particularly good one. If, however, there is any implication that Kant's list is absolute, my camel must respectfully balk. Kant's palette may contain subtler colors than does the ancient pagan's, yours, or mine; and, Kant's skill as an artist

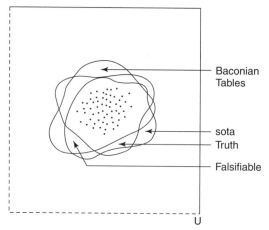

Figure 41 Definitions of Science

may be more developed, but on close inspection, his finished world is found, just like ours, to consist of individual dabs of heuristic. The only categorical imperative (and it is not an ethical, but only a compelling, one) is to structure your world according to your personal sota.

Fuzzy Subsets

Partitioning the overall set of heuristics into these smaller subsotas is a valuable heuristic with the caution that the operation is neither well defined nor absolute. Subsotas have fuzzy boundaries. The use of the strange word *fuzzy* is meant to imply that the boundary separating those heuristics that are within a set and those without is not precisely determined in an absolute way. The person who does the partitioning plays a decisive role in deciding which heuristics to include. Since differences of opinion are inevitable, no exact, well-determined boundary is possible. The fuzzy subsets *science, common sense,* and *man* will make this point.

Different scientists will include different heuristics in the definition of science. The most radical scientists go so far as to assert that the sotas of the natural sciences and the social sciences have no overlap at all. More reasonably, the heuristics of comprehensibility and objectivity are routinely included in the sota of science, but wide differences of opinion exist as to what else should be included. In Figure 41 three slightly different definitions of science are carved out of the overall sota, each represented as a closed curve. These definitions of science share some heuristics, but each contains heuristics unique to itself. For example, the subset that Francis Bacon would have used to define science contains the heuristic of setting up the tables he proposed as an aid to scientific induction. This heuristic would not be included in Karl Popper's definition of science. Popper suggests that an important characteristic of a proper scientific theory is that it should be falsifiable. Other definitions suggest that science converges to truth or that it is nothing more than a set of heuristics or sota. The person who defines science sets up his own eccentric partition, and the present state of the art of the concept *science* is, therefore, best represented as a subset with fuzzy boundaries.

Occasionally the boundary that defines a concept is relatively definite. Often large numbers of people, sometimes entire cultures, and at times nearly all humans agree on the boundary of a subsota. The existence of God, the external world, other minds, truth, and reality are examples of heuristics whose importance most people hold in common. To this overlap, to this commonality of opinion we should give the name *common sense* rather than to "what is evident by the natural light of reason," as some have proposed. Reason is already too small a subset to be convincingly further subdivided into common and uncommon sense. *Common* meaning "common among a group of people" is a more commonsense heuristic than anything having to do with the young, restricted, and minority view called reason. Common sense is simply a cluster of heuristics that a common group of people is compelled by their sota to believe are good heuristics.

Man is also a good example of a fuzzy sota. He is defined by a subset of all heuristics and defined differently depending upon who decides which heuristics to include in this subset. At one time or another, opinions as to which heuristics are essential for a definition of man have emphasized man's soul, mind, and body. He is the rational animal or the featherless biped or the image of God or the animal with language ability—it is for you to choose. For others, man is only an illusion created by the heuristics of division, categorization, and space. Ultimately, man is man-made by the person choosing which heuristics to include in the subsota he uses to define man. Taken as a whole, the concept *man* is a fuzzy subset of the overall set of all heuristics.

Personal sotas

Among all of the competing subsotas that make up the fuzzy sota called man, none is more important, more distinct, and more emphatic than the one that contains the heuristic "I am defining myself," the simpler heuristic "I exist," or, in the form I prefer because it avoids the prejudice of English grammar, "I." Many philosophers maintain that this sota is immediately and intuitively known, existentialists construct their systems with respect to it, and most people are aware of and puzzled by its hold over them. Although we can question the existence of other sotas, we do not even know how to go about questioning our own. The set of heuristics you use to define yourself is the querencia where you feel the most secure, the sota that you will defend to the death.

An extension of the set of heuristics we use to define ourselves is the one that defines our view of the world. We talk, we think, we know about the world—we structure the world in terms of concepts or subsotas. The collection of subsotas each of us uses is the palette with which he creates his world. The ancient pagans painted their world in terms of their Gods: Zeus, Mercury, and Mars. We, thinking our art less primitive, develop ours using the concepts: truth, reality, and absolute. These are not just two different ways of looking at the same hypothesized external world, but two different worlds, two completely different sotas. To use our word *heuristic* and paraphrase an often quoted parable given to us in a slightly different context by Otto Neurath in "Protokollsätze," *Erkenntnis*, 3, 1932, "We are as mariners at sea in a boat built of heuristics. All we can do is rearrange and improve these heuristics while trying to remain afloat."

No longer do we even ride on the back of a turtle. That metaphor is too concrete. Now each of us strives to stay afloat in a boat of our own making. The heuristics we use are unique to each of us and inaccessible to others. The hermit high on a mountain in India, the banker in a skyscraper in New York, the Australian aborigine, and the Pygmy in Africa define their worlds in terms of different sotas. The difficulty comes when we try to give equal validity to sotas that are so distinct from our own. We cannot help but believe that our Western, scientific-based sota is an absolutely superior description of man in his world than is the Eastern, Vedic-based sota of the Indian hermit who remains in seclusion for decades atop a cold, wintry mountain. How can we doubt that the wave function of our culture is not better than the maya of his? We are, however, on one side of a mirror, and he is on the other. As we raise our fist to insist that we are right, he raises his hand to insist that he is right. As we claim that he has lost touch with the real world, he claims that we have lost touch with our true nature, that we have forgotten the meaning of maya, that we are taking the concepts in our sotas for the realities of nature instead of realizing that they are concepts of our mind. As we use our heuristics to make our case, he uses his heuristics to make his case. *Each of us speaks with the accent of his own sota.*

Only by realizing that we have been captured by our sota and that he has been captured by his can we grow to tolerate the differences, recognize the separate validity of the two subsotas as subsotas, and achieve harmony in the two points of view. Since one sota cannot be completely translated into another, each person remains fundamentally isolated. To this sense of uniqueness, inaccessibility, alienation, and loneliness some philosophers attribute the absurd nature of the human condition. A person's sota is a Procrustean bed, and he has no choice but to lie down in it.

The trail for finding universal method has led into unexpected territory. Discovering the meaning of the heuristic has been the most difficult and time-consuming part of our hunt. Using the engineer's rule of thumb as an initial guide, we arrived at last near our first destination, the heuristic. Although we are not completely satisfied with our findings because a feeling for the heuristic depends on the heuristic itself, if we remember that this discussion is only intended as a first cut, our progress should be judged satisfactory. Next, in quick succession we found the meaning of the overall sota, the fuzzy set of heuristics called man, and now the further subdivision each of us uses to define ourself and our world. What began as a pleasurable excursion in engineering has quickly become a more strenuous adventure in philosophy. Having come such a long way, we are undoubtedly tired, and so I propose a brief rest to view the surroundings. For a break, let's consider two characteristics of the sota we use to structure our worlds: its compelling nature and its incoherence. They will show the way to the *ataraxia* of the engineer we all seek. Once we are thus refreshed, we will return to the trail to investigate one special partition of the overall sota called *method*.

Compelling Nature of Personal sota

The vicelike grip the set of heuristics a person uses to structure the world has over his every action, thought, and mood is the first characteristic we must address. Many philosophers now agree that skepticism (and now, perhaps, they would include the heuristic view) is impossible to refute but still impossible to believe. They agree with Hume that "Nature, by an absolute and uncontrollable necessity, has determined us to judge as well as to breathe and feel" and with George Santayana that "belief in the existence of anything, including myself, is something radically incapable of proof, and resting, like all belief, on some irrational persuasion or prompting of life." Try as you may, you cannot *will to doubt.* You may at last agree that doubt about everything is the best of philosophies if you could only maintain that doubt, but you find that Nature is too strong. Sitting opposite me, you glance down to see the book that you hold in your hand; you smell the glue of its binding; and your head begins to reel. You turn on the television set; you sit down to dinner; you rush next door where some friends are playing billiards; and the real world floods in. You simply cannot disbelieve. I mischievously misquote Santayana. The real world must be intelligible or you must tear your hair and go mad. What is the nature of this Nature? It is nothing but the effect of a person's sota as it both limits and compels his actions.

To demonstrate this point and to prepare for the *ataraxia,* or peace of mind, of the engineer, I now have a request to make that is somewhat unorthodox and incongruous with the seriousness of purpose of this dissertation. The heuristics of formal writing dictate that the author remains at a distance from his reader if he wants to be taken seriously. I acknowledged this rule of thumb in the Introduction, warned that I was required to violate it in our discussion, and have repeatedly done so to prepare for this moment. Now I must go one step further and ask you to play an admittedly childish game. To demonstrate the way each of us is limited and compelled by his personal sota, I have chosen an incisive amusement and ask you to join me.

My Guillotine

We are now to consider a variety of people each of whom is confronted with the choice of removing his head from a guillotine or leaving it there to be chopped off. We are then to predict which choice each will make. Please decide whether or not to play my game and seal your choice by putting a checkmark in the following blank if you decide to play _____. An odd request, I grant, but I assure you that it is essential to my purpose.

Now that you have made your decision, let me set the stage for our game and introduce the players. Various people are to find themselves with their head unrestrained in the mouth of a guillotine at ten minutes until twelve o'clock noon. Each is free to remove his head if he pleases, but, in any event, at twelve o'clock sharp the blade will fall. The cast of characters includes the following condemned persons:

1. Pyrrho
2. A traditional skeptic
3. Hume

4. An engineer

5. You, yourself

If you have agreed to play, please review this list and put a small checkmark by the names of those who you think will lose their heads. Do not forget to include yourself and pretend that your neck is also on the block. I'm sure you would remove your head—but can you give a compelling reason why? I am willing to share my scoring with you, but at twelve o'clock all matters of opinion are rendered moot.

First, Pyrrho would surely lose his head. History records that the disciples of Pyrrho followed their master everywhere to keep him from walking off a precipice or being run over by a cart because he trusted nothing more than anything else. My guillotine would not bother him.

Second, the traditional skeptic stands a good chance of losing his head, although somewhat less surely. The skeptic claims to suspend judgment on all questions because equally good arguments can be offered on both sides. As a result he would find himself in the position of Buridan's hapless ass. As you recall, this poor animal was exactly as thirsty as he was hungry and chanced to find himself exactly halfway between food and water. Unable to choose which way to turn, he died in plain view of what he needed to live. The skeptic is equally asinine and would hesitate between believing and not believing his senses when they reported that he was in peril. As he tarries with his neck on the block, the clock strikes twelve. Any reservation I may have about predicting the demise of the skeptic is because of the escape clause some use to mitigate their position. As was noted earlier, Montaigne feels that ultimately all we can do is live according to nature and custom and accept what God chooses to reveal to us. Hume reminds us that "even the most extreme ancient skeptics admitted that we must live; and since reason cannot tell us how to live, they recommended that we follow nature." Therefore, I hesitate in being too dogmatic about what the skeptic would do until we can decide what brand of skeptic we have in mind.

Guessing the life expectancy of Hume, himself, is a little less difficult. He felt that no logical necessity connected cause and effect or justified induction. Logically, then, Hume has no reason to fear the falling blade. Psychologically, however, he did try to justify action by calling on *habit* and *custom*. Since Hume could not have had direct experience with a guillotine, what custom or habit he evokes in the case of my guillotine is necessarily secondhand. That is, Hume is left with one of two indirect appeals to save his neck. First he could turn to habit established by past experience with sharp, heavy, falling objects that are critically unlike a guillotine and by some as yet unspecified heuristic (certainly not induction, for by what habit could he have made this *particular* inductive leap before?) make the transfer to the case at hand. Second he could use habit with his past experience with the truthfulness of people in reporting the effect of a guillotine to justify his faith in their report of the consequences of a guillotine now that their opinions really matter. The first contains a step on the way to truth that is so vague as to be unintelligible; the second suggests that Hume led a very sheltered life. With the many available counterexamples, trusting one's life exclusively to the habit of believing in the veracity of man is liable to produce a surprise Hume would not be around to appreciate.

At last we come to the engineer. As you have surely guessed, I would have him remove his head. His choice is based on, constrained by, and compelled by his sota. From the engineer's point of view, only two possibilities or heuristics exist: either to remove his head or not to remove his head. He admits that the future may bring a third possibility, but for the present he is restricted to one of two choices. At ten minutes until twelve his sota presents him with a forced choice between the two, and at twelve o'clock he will have acted on one of them. He does not seek "the true thing to do," but "the best thing to do." He must act upon a heuristic; the only issue is upon which one. He must either believe that the future will be like the past or not. He must either believe that he should apply past experience with guillotinelike situations or not. He must either remove his head or not. Only these contradictory opinions are available. At ten minutes until twelve his sota presents him with a forced choice between the alternatives, and at twelve o'clock he will have chosen between them. The sota of most engineers will compel them to choose the heuristic *Remove your head immediately.* The important point is that as heuristics both alternatives are equal. Only from the privileged perspective of the specific engineer does one or the other become a good heuristic.

This brings us to *you.* If you have accompanied me this far, you have proven that you, like the engineer, would surely remove your head and be saved. Will you let me show you how?

Stop and think. You have either agreed to play my little game and put a mark in the blank above or you have not. You, like the other condemned men confronted with the guillotine, are under a restricted choice based upon the alternatives in your sota. The future may bring a third possibility in addition to playing or not playing my game much as Gödel added a third alternative for the truth value of a proposition that was unknown before him. For the present, my sota, the engineer's sota, and, I am sure, yours compel us to believe that the blank in this book now either contains a checkmark or it does not. We cannot even conceive of an additional alternative, although we cannot provide a convincing, absolute reason why we cannot. *A person's sota fundamentally constrains and compels his actions.*

This example is not contrived or its bite the result of a moment's inattention on your part. It will work a second time. If you now erase your previous checkmark if it was there and reread the choices, do you doubt that you still have a restricted choice? Feeling the cold steel of the guillotine blade on the back of your neck, perhaps you will struggle. "The checkmarks were only signs chosen at random and were in no way correlated with my decision to play your game," you object. Good point, but actually I have only marginal interest in whether I can tell if you have agreed to play the game. That depends on the constraints imposed by the heuristics in my sota on me, not those in yours on you. Shall we try again? This time leave me out of the analysis, do not use checkmarks, and do not bother to tell me your answer. Decide for yourself whether or not to go back and replay the game. Try to convince yourself that there is an additional alternative besides deciding and not deciding to play and act on this third choice. Remember at twelve o'clock you will have made your decision and the blade will fall.

You have your scorecard, and perhaps it is different from mine. This is to be expected because our sotas are different. But at twelve o'clock, Pyrrho, the skeptic, Hume, the engineer, and you, yourself, will either be walking around with his head or lying around without it, and we could compare our results.

My guillotine exploited a heuristic firmly entrenched in the sotas of most people in the Western world, *the law of contradiction*, to demonstrate the limitations of a person's sota and its compelling nature. The challenge was to remove your head from the guillotine or not, to replay the game or not, and to decide to replay the game or not. Other heuristics, such as time, could have been chosen to make similar points. Despite the conclusions of the modern physicist that the absolute distinction between the past, present, and future is no longer valid, try to convince yourself that for some observers you read the sentence you are now reading *before* you placed your checkmark on the previous page. We live in privileged times. The same examples might not work for other eras, cultures, or places. The law of contradiction, time, space, reality, causality, and so forth— what is the origin of this Nature that is so compelling to us? Often it is nothing more than an accident of birth, but that fact does not weaken its hold over us. Most people who endorse Western philosophy were born into the Western philosophical tradition. Most people who endorse Eastern philosophy were born into the Eastern philosophical tradition. How else are we to explain the extraordinarily high correlation coefficient between underlying worldview and where a person was born? Speaking of which, what is the correlation coefficient between where a person was born and the religion he chooses to regulate the most sacred part of his being? *A person's sota is the one true dictator who cannot even theoretically be overthrown by the one oppressed.* We will see that the peace of the engineer results from recognizing and accepting this fact.

Incoherence of Personal sota

The set of heuristics a person uses to structure his world does not behave as a single, coherent unit. Different, often contradictory, heuristics are called into action depending on the circumstances. I witnessed a striking example of this during a conference in San Antonio, Texas with a colleague, a theoretical physicist.[*] When I casually asked my friend if he thought my son could have been on the moon, without hesitation, he answered "no." Somewhat later in the evening, I told him that I needed some information about the relative probability density function $\int \Psi \Psi^* d\tau$ for a book that I was writing. Under further questioning he admitted that this symbol represented the probability of finding an object, such as my son, at various locations in space, and that it had a small, but positive, value on the moon. This answer was, of course, in direct contradiction to his earlier stance. To confirm this experience, I repeated the experiment at a dinner party of another close friend, who was also a theoretical physicist,[†] with the same results. This incoherence is not carelessness of thought. It is consistent with the philosophy of Wittgenstein, who felt that a statement could have two meanings depending on who uttered it. Wittgenstein admitted

---

[*]Dr. Art de Charme.
[†]Dr. Wendell Horton.

that he tended to challenge the knowledge claim "This is my hand" when made by a philosopher, but not to question it when a person makes the statement as a casual observation in everyday life. The theoretical physicist uses two different heuristics to answer the same question depending on the external sota that confronts him.

Under certain circumstances the lack of coherence in a person's sota becomes intolerable. If we become obsessed with these inconsistencies and forget that they, and the notion of an inconsistency itself, are only heuristics, we find that we have painted ourselves into a corner. For example, a feeling of perplexity develops when we consider the power of God. After assuring his son that God can do anything, a father is at a loss when his son then asks, "Can He make a stone so large He cannot roll it?" This paradox is often dismissed either by arguing that if God can do anything He can figure a way out of this situation. Calling on divine powers to relieve our frustration does not work with the related question, "What happens when an irresistible force meets an immovable object?" By setting up the dichotomies "can do anything" versus "cannot roll a stone" and "irresistible force" versus "immovable object," we weave a net of heuristics, forget that we have done so, and panic when we become ensnared. When concepts such as *irresistible*, *immovable*, and *cannot* are taken for more than heuristics, we find ourselves in complexities of our own creation.

The perplexities created by the stone God can't roll and the irresistible force merit only a moment's reflection, to be dismissed until we chance upon them once again in a book. They are considered quaint, somewhat artificial problems by most Westerners. The discomfort they cause is, however, a readily accessible model for the deeper, more profound malaise that results from the incoherence in a person's sota and from forgetting that *All; heuristic*. Specific examples will demonstrate this point.

Simply stated, determinism is the general philosophical thesis that for everything that ever happens there are conditions such that, given them, nothing else could happen. In contrast, free will is the doctrine that the will of an individual can and does determine some of his acts. Is man free or is he a machine? The most famous philosophers and religious writers of all ages have examined, modified, and delineated the various aspects of this problem. Despite this mountain of analysis, each generation is struck by the problem anew. As you read this, your sota probably contains an unstable blend of the two positions. If you are bothered by this situation to any degree, you have also painted yourself into a corner. Both determinism and free will and the feeling that there is a problem at all are heuristics. Our sotas contain a heuristic labeled "free will," another labeled "determinism," a third labeled "not both at the same time," and a fourth labeled "no other possibilities can exist." If the first two appear separately in appropriate contexts, we register no concern. Unfortunately, history has too often thrust all of them onto the stage at the same time, but never allowed us to see who might be waiting in the wings. This does not mean, of course, that it is not a good heuristic to think about the problem or, if you insist in removing conflicts in your sota, to try to reconcile these conflicting ideas. But it does insist that it is the supreme egotism to think that our man-made sotas and the inconsistencies they contain should have absolute cosmic import.

Another example of painting ourselves into a corner is the mind-body problem, sometimes called metaphysical dualism. How can a mind, which is not spatial, cause a particle of matter to move; or how can matter, which can only move from one place to the other, produce an idea in the mind? Various solutions have been offered to this recurring dilemma: God wills it (*Deux ex machine*); mind and body evolve on a parallel track (Leibnitz); denial that matter and mind are different (neutral monism); denial that matter exists (idealism); and denial that mind exists (behaviorism). Write mind, body, dualism, move, and idea and the horns of the mind-body problem disappear.

I will not bother you by considering the age-old controversy over the objectivity or subjectivity of good and evil, for now you can tame it in much the same way. Instead I pass on to such ultimate questions as "Why do we exist?", "Why does the world exist?", "Why is there something rather than nothing?", or simply "Why?" These questions are the final downfall of all philosophers. Some have frankly admitted that their philosophies do not pretend to provide the ultimate grounds for the answer to these questions; some have become lyrical; some have become mystical; some have turned to religion; some have bowed their heads in silence. The human sota has fashioned this tangle more effectively than King Gordius of Phrygia did his. I weld against this knot the sword *All; heuristic* and with a single, decisive stroke reduce "Why?" to "Why?" God's stone, the irresistible force, the Japanese kōan, determinism versus free will, the mind-body problem, and the simple word *why?*—each of these examples will only make us feel uncomfortable if we take our heuristics and certain combinations of them too seriously and, in effect, paint ourselves into a corner.

Rules of Judgment and Implementation

Little needs to be added to draw our attention to the appropriate rules for judging an individual, the rule for implementation, the manner in which the sota of humanity evolves, or how it is transmitted from one generation to the next. We reason by analogy to what has come before.

If the heuristics in every culture, every region of the world, and every age were combined, it is against this huge, comprehensive set that each of us should be judged.

> The fundamental *Rule of Judgment* is to evaluate an individual and his actions against the sota that defines best practice at the time he is to be judged.

As with the sota that represents best engineering practice, the one that represents best human practice is unknown to any single human, and we are forced to call on a heuristically determined subset of humans for advice. Likewise I find I have a sufficient number of rules to live by if I observe the rule:

> The *Rule of Implementation* as a human born with intelligence is in every instance to choose the best heuristic for use from what my personal sota takes to be the sota representing best human practice at the time I am required to choose.

Similar to the case of engineering, best practice evolves in time and we can only hope that the heuristics we select are near optimum. In an analogous way, the heuristics we know are transmitted from one generation to the next by learning at our grandmother's knee, from books, in schools, and from the concrete sota into which each of us is born. *To be human is to be an engineer.*

Engineer's Ataraxia

The world of the dogmatist is overpopulated, arbitrary, and according to one expert nothing is so absurd but that it may be found there. I would not live in it. The world of the skeptic is morose, filled with lethargy, and according to another expert as barren as it is irrefutable. If anything, I find this even worse. Where can we find the promised peace, the *ataraxia,* we all so desperately desire? Through the desert of skepticism, the kōan *All; heuristic* leads the way to the promised land. All heuristics are equal as heuristics, but from the vantage point of your sota some are better than others. We are compelled to use heuristics—the only question is which ones we will use. I release a ball one hundred times and it falls. I now open my hand for the one hundred and first time, and you are asked what will happen. Surely you feel that it will fall. Why? Falling and not falling are both heuristics. You are constrained to choose between them. There can be no logical justification for believing the ball will fall, and depending on an explanatory fiction such as habit or custom to justify belief is unsatisfactory. The connection between opening the hand and the ball falling is not logical, not necessary, not habit, not custom, not some irrational persuasion, not the prompting of life, but heuristic. That it will fall is a good heuristic in most people's sota. All choice must be forced back to the limitations of the specific, concrete sota of the person in question. This strategy is convincing by being immediate and intimate instead of abstract and susceptible to attack. The peace of the engineer derives from a tolerance for all heuristics as heuristics and a preference for what he is compelled to accept as the best heuristics. A person's sota defines his actions, thoughts, and moods. If it includes the good, he will do the good.

Additional examples will clarify the crucial heuristic needed to achieve the peace of the engineer. "Knowledge is possible" and "knowledge is not possible" are both heuristics with no intrinsic reason to prefer one to the other. Compelled to choose between these alternatives, most engineers, indeed most philosophers and laypersons, will choose the first. "Suspending judgment" and "not suspending judgment" are both heuristics of equivalent importance as heuristics. If the skeptic chooses to suspend judgment, he has simply made a choice of heuristic based on his sota on a par with, but no different from, the engineer who chooses to judge. He has no cause for pride that he has done anything special. "God is dead" and "God is not dead" are heuristics in the overall sota. The individual sota assigns to one or the other the label *good*— the Pope's sota compels the first; Nietzsche's, the second. Because the sota of the engineer, unlike those of many philosophers, dictates a world that is straightforward, concrete, and devoid of crucial obscurities, the engineer is able to raise science, truth, absolute, and reality from the ashes to assume their traditional importance by accepting belief in them as a better heuristic than denying them.

The heuristic *compel* acknowledges the dictatorial power of a sota. A person can no more will to believe a heuristic that he does not believe and have it compel his actions than he can increase his height or eradicate his distaste for endives. The compelling nature of the set of heuristics a person uses to structure his world is the nature of that Nature that is so strong.

The engineer's peace is elusive only because maintaining the basic premise of this discussion for any length of time is difficult. The philosopher who tries to do so is as one who sits on a large rock at the exact summit of a steep hill. In this position, the system is in equilibrium, all directions are equivalent, and the rock will stay where it is forever. As long as he sits still, the philosopher is unperturbed, at rest, and untroubled by mental or emotional disquiet. If, however, he shifts position ever so slightly, both he and the rock will go careening down the hill out of control in one direction or the other. Although the ride down may be exhilarating, once at the bottom, the philosopher can only pick himself up, dust himself off, and wonder how he arrived in the briar patch of idealism, empiricism, pragmatism. Make no mistake: Skepticism is also at the bottom of the hill, but rather than enjoy the ride down, the skeptic digs in his heels, protests, and claims to doubt that he is going for a ride. After each trip down, the hard part is pushing the rock back up the hill to regain the original position. Maintaining the rock on the pinnacle where *All; heuristic* is the modern Sisyphean task.

A DISCOURSE ON METHOD

Although the title of this book promised a discussion of the method, it is over 85 percent finished and only token mention has been made of this important concept. The break to consider the compelling and incoherent nature of the sota we each use to structure our world finished, we must now return to the trail to rectify this slight and tease a statement of method from the background we have so carefully prepared. This is the last objective of Part IV.

Method, like man, like engineer, and like American engineer, is a fuzzy subset chosen from the overall sota. As with all fuzzy sets the exact heuristics it contains depend on the person doing the partitioning and are liable to change in time. As I now seek method, I return to the subset that I have found compelling since the spring of 1965.

As mentioned when we first encountered the word *method* back when our only concern was the engineering method, *method* is thought to have first appeared in the dialogue *Phaedrus* by Plato and to have come from the two Greek roots meaning "along" and "way."* Therefore, method refers to the series of steps that are taken along the way to a given end. It tells what you must do to achieve a specific goal.

    1. I want to do engineering design; I seek the method of the engineer.
      • Do I want change?

---

* μέ θοδοs, "a following after, a scientific treatise, method, system, artifice, ruse"; Prof. Paul Woodruff.

- Do I want the best change?
- Do I want the best change in a complex, uncertain situation?
- Do I want the best change in a complex, uncertain situation within the available resources?

In sum, what do I do?

2. I want to do scientific research; I seek the method of the scientist.
    - Must I be objective?
    - Must my results be reproducible?
    - Must I assume the world is comprehensible?
    - Must my theories be falsifiable?

    In sum, what do I do?

3. I want to write a book; I seek the method of the writer.
    - Should I maintain a constant rhetorical profile?
    - Should I address the reader directly?
    - Should I include an ablative absolute, an *ad rem*, or a fragment?
    - Should a sentence begin with *and*?

    In sum, what do I do?

4. I want to paint a picture; I seek the method of the artist.
    - Do I apply gesso and make a preliminary charcoal sketch?
    - Do I develop the picture as a whole or concentrate my effort on one part of the canvas at a time?
    - Do I reaffirm my line periodically?
    - Do I appeal to the emotion or intellect?

    In sum, what do I do?

Common sense requires the fuzzy subset called method to contain heuristics that answer the fundamental question, "What do I do?" Method is the sota that enjoins.

Since Plato, the discussion of method has often centered around two heuristics: *analysis*, the reduction of a complex object to simpler ones, and *synthesis*, the reverse process. These two ideas have appeared under a variety of different names: analysis as *resolutio, resolutiva, analytica, methodo risolutivo*, reduction, resolution, division, and deduction; synthesis as *compositio, compositiva, synthetica, methodo compositivo*, composition, combination, and induction. Of course, nuances separate the terms in each of these two series, but for present purposes it is sufficient to say that the historical answer to the fundamental question, "What heuristics define the fuzzy subset called method?" is "analyze and synthesize."

Many questions have arisen concerning this answer. What is the relative importance of the analytic and synthetic aspects? Aristotle emphasized deduction in his great work, the *Organum*, often translated as tool, instrument, or method, and Bacon chose induction in creating the new method of science in his equally famous work, *Novum Organum*. Are analysis and synthesis parts of one method or two separate methods? According to Locke, both aspects taken together form the essence of method, but according to Newton two separate methods exist with the analytic always preceding the synthetic. The case of Descartes is in dispute. Some of his followers claim a single Cartesian method with two aspects; others claim that Descartes distinguished between two methods, those of analysis and synthesis. Rather than try to resolve these differences of opinion at the present time, a more interesting question is, "Is there one method appropriate to all of the sciences or a different method for each?" Aristotle, Saint Thomas, and Galileo all agreed that each science has its own method, as opposed to Bacon, Leibnitz, and Condillac, who felt that only one scientific method exists. Even if we assume that one method for all sciences exists, is there a universal method for all realms of knowledge? Many authors attribute to Newton a method for attaining knowledge in any field, and Descartes makes the specific claim that his method is universal and applicable to all subject matter. From the seed planted by Plato, a field of questions has grown. Of all of them, only one interests me: To the question "Is there a universal method?" I answer "yes." The remainder of this part of our discussion is to explain what it is.

Method of Descartes

Any discussion of a universal method must begin with Descartes because he was one of the first and certainly the best-known philosopher to have made the specific claim of having found one. For Descartes, man is essentially a rational being and an account of reason in its pursuit of knowledge is at the same time an account of method. As with many of his predecessors, the central feature of the Cartesian method is analysis and synthesis. In answer to the question "What do I do?" Descartes gives the following counsel. Reduce whatever is complex in experience to its component parts until a stage is reached in which the objects are known to be true in themselves or, as Descartes would say, until our knowledge of them is absolutely *clear* and *distinct*. Then build on that. All Descartes felt he needed to construct the edifice of knowledge was one such *clear* and *distinct* idea, and he claims to have found it in the expression "*Cogito, ergo sum*" (I think, therefore I am). He asserted that our knowledge of this fact is absolutely trustworthy. Any attempt to refute it is an example of thinking and hence confirms it. Borrowing only the first Latin word for simplicity, the first goal in Descartes' method is to reduce whatever is complex to the *cogito* that is then taken to be intuitively grasped as true. The second step is to construct the rest of knowledge on this firm, sure basis. Descartes' method is summed up in the four famous rules in the *Discours de la Méthode* and more fully developed in the twenty-one rules in his posthumously published *Regulae ad Directionem Ingenii*. All reasoning is conceived as a series of intuitive steps based not on the rules of formal logic but on the following practical injunctions. In Descartes' words:

> The first rule was never to accept anything as true unless I recognized it to be evidently such: that is, carefully to avoid precipitation and prejudgment, and to include nothing in my conclusions unless it presented itself so clearly and distinctly to my mind that there was no occasion to doubt it.

> The second was to divide each of the difficulties which I encountered into as many parts as possible, and as might be required for an easier solution.

The third was to think in an orderly fashion, beginning with the things which were the simplest and easiest to understand, and gradually and by degree reaching toward more complex knowledge, even treating as though ordered materials which were not necessarily so.

The last was always to make enumerations so complete, and reviews so general, that I would be certain that nothing was omitted.

With these four rules, Descartes defines his universal method.

Problems with Descartes' Method

I agree that all ancient authors should be left in undisputed possession of their honors—Descartes all the more so since he was the inspiration of this work. I have no taste to play the critic for if nothing else, the heuristic philosophy is a benign one, allowing to each philosopher his own eccentric view or set of heuristics. For each scholar his definition of method is necessarily the best and most compelling, which is the most that can be said of any idea. Therefore, I will not review past efforts to refute the Cartesian system. Still I do need to sort out a few difficulties that I personally have with Descartes' position before I can state my own definition of the universal method. In particular, I would expect the universal method to have some important characteristics that seem to be lacking in the Cartesian system. It would seem to me that a universal method should be universal, comprehensive, as nearly independent of a prior philosophical commitment as possible, and self-sufficient. Since the Cartesian universalization of method confuses me on each of these accounts, closer consideration of these problems is desirable.

Universal

Whatever characteristics the fuzzy subset labeled universal method must have, at the very least, it should be—well—universal. Universal method must denote the actual method used by everyone: the scientist, the philosopher, the engineer, but also the layperson, the madperson, the hermit, and the aesthete. A method can hardly claim to be universal if it is not used by the Chinese, the Indians, and the natives of Pago Pago, or by the young and the old, and by the rational and the irrational. Whether lofty or inconsequential, at every moment everyone *does* something. He makes decisions, cooks, sings, loves, or just sits still. What he does must conform to the demands of any method that claims to be universal. Universal method tells what *is* done, not what *should* be done. Again, in *Discours de la Méthode*, Descartes specifically warns that two sorts of individuals should not use his method: "Those who think themselves more able than they really are, and so make precipitate judgments and do not have enough patience to think matters through thoroughly" and "Those who have enough sense or modesty to realize that they are less able to distinguish the true from the false than are others, and so should be satisfied to follow the opinion of these others than to search for better ones themselves." That you should have a choice as to whether to use universal method or not seems a bit odd, doesn't it? Let me insist: What is at issue with the universal method is not what you should do, but what in fact you always do. It may be one thing to philosophize and another to live, but it most certainly is not one thing to philosophize about a universal method and another to follow it. To be universal by any common-sense definition of universal, a method must encompass what everyone must do at all times.

Comprehensive

Common sense also suggests that a universal method must include not just all methods that are actually used, but all conceivable methods. That is, it should include not just all good methods, not just all well-defined methods, not just all philosophical, artistic, and scientific methods, not just all methods for seeking true knowledge, but all methods without exception. It must include all conceivable answers to the fundamental question "What do I do?" That is, a universal method should be comprehensive. Consider the method: Perceive everything as a Gestalt. My claim is not that this method of the Gestalt psychologists is necessarily a good method, but only that it is a method. The Cartesian method, being reductionist, runs counter to the advice it gives. If the four rules of Descartes are to be taken as universal, then a person who perceives one of the famous Gestalt figures is not doing anything; he is not following a method. Or consider another even more extreme example of a method: Never use the Cartesian method. Common sense accepts this as a method. It tells what to do in the form of a clear injunction, but it cannot possibly be Cartesian. Many particular methods exist; universal method carries an extra burden—it should include all of them as special cases.

Prior Philosophical Commitment

Ideally the statement of a method claiming to be universal should be free of all prior philosophical commitments. If it is not, we risk losing those who reject that upon which the statement of universal method depends. We should also wonder by what method we are to come to know these prior commitments if they stand before and apart from a universal method. Since a complete lack of prior commitment seems impossible at the present time, those commitments we do make should be as few in number and as general in scope as possible. At a minimum, they should include and apply to themselves and to the universal method itself. The Cartesian universalization of method requires philosophical commitments that seem troublesome in these regards.

As stated before, following a method is doing something. It is exhibiting behavior. A description of a universal method should not depend on extraneous assumptions concerning the nature of the behaving organism, but focus on its behavior. Descartes' system depends on the definition of man as a rational animal because reason and method are specifically equated. Rejection of his definition of universal method follows logically from rejection of his definition of man. While it seems reasonable to deny the existence of *rational* method if no rational organisms exist, it does not seem reasonable to deny the existence of *universal* method on this same basis. A description of universal method should attend to what man does, not what he is.

Too limited a prior philosophical view raises problems in understanding what a likely candidate for a universal method even means. When words like *rational, order, clear, distinct, simplest, parts,* and *certain* appear in a definition of the universal method, the skeptic is bound to ask by what method he is to know what *they* mean. For example, to justify the *cogito* as the sure foundation upon which to build by calling it *clear* and *distinct* requires another method to justify the notions of clear and distinct. As suggested earlier, all philosophy changes subtly if its unjustified concepts are underscored to indicate that they are only heuristics. Descartes' four rules are no exception.

If a universal method cannot be completely free from any prior philosophical commitments, at the least, we must insist that the commitments it does make should be sufficiently broad to apply to all definitions of the behaving organism and include a sufficient mechanism for expressing itself.

Self-Sufficient

A universal method must not depend on another method for its operation. Any description that includes more than one rule would seem to need rules to specify the relationship between all of the rules it has. Particularly critical is a method for choosing which rule to use when. A method that is complex requires a meta-method for its operation, a method to articulate the parts of its complexity. If this new method is exterior, the original set is hardly universal. Therefore, a complex, universal method must be self-sufficient and include internally all information necessary for its successful operation. The rules of Descartes are independent and have little clear, internal instructions for their mutual interactions. Worse still, there is no internal rule, no internal guarantee, no internal feeling that the set is even complete. Descartes' four rules have no sense of closure about them. The simplest way to avoid the problem of articulation between separate rules is to have only one. An effort to encapsulate the rules of the Cartesian method into one rule by defining it as the descent from the complex to the simple until an intuitively sure point is reached and then the ascent to develop the more complex reveals other hidden demands for auxiliary methods. Does everybody necessarily reach the same fundamental object known clearly and distinctly? If so, by what method do we know that; if not, by what method do we relate these separate fundamentals or assure that all of them will develop into the same true knowledge? Any method worthy of the title *universal* must not depend on additional injunctions to assure its operation.

Since I am unable to resolve my confusions over the universality, comprehensiveness, independence, and self-sufficiency of Descartes' method, I feel obliged to describe the universal method I find compelling.

Universale Organum

The heuristics for getting a clean shave with a straight razor suggest that 90 percent of the effort should be invested in the preparation and only 10 percent in the final execution.* If the lather is properly prepared, the skin and

---

*Dr. Robley Evans, circa 1963.

beard are softened, and the razor is sharpened, all time-consuming tasks, the face is shaved in minimum time. Like shaving, this discussion has allocated a majority of its resources to preparation. As a result, the definition of the universal method may follow quickly and almost anticlimactically. We have discovered that to give an account of a method, it is sufficient to give a description of what a person must do when he is following that method. Method is the injunction that answers the question "What do I do?"; universal method is the injunction that answers the question "What do I always do?" Making the single philosophical commitment *All; heuristic*, I have found that I can make do with only one rule provided I make a firm and unalterable resolution not to violate it even in a single instance. The *universale organum** I find compelling is the

HEURISTIC: Use heuristics.

I must admit that I am relieved that I can make do with only this one rule. If I needed more than one, I would always worry that I might need yet an additional rule, not included in the original set, to tell me which of my original rules to use at each moment.

Make no mistake: This definition of the universal method is not meant to imply that humans just use heuristics from time to time to aid in their work, as might be said of a mathematician. My thesis is that what all humans do at all times and the use of heuristics is an absolute identity. For me, the grand sweep of this generalization from engineering method is truly exhilarating.

The *universale organum* does not suffer the other defects I have attributed to the Cartesian universalization of method. To begin with, the method is truly universal. Philosopher, scientist, engineer, layperson, and child are all constrained by their respective sotas to follow the injunction to *Use heuristics*, although the particular set from which they choose may differ. In other words, the overlap of the fuzzy sotas—$\text{sota}|_{\text{phil.}}$, $\text{sota}|_{\text{sci.}}$, $\text{sota}|_{\text{eng.}}$, and so on—contains, at the very least, the fundamental heuristic that defines the universal method, *Use heuristics*.

On the other hand, all methods are subsets of the *universale organum*. The Cartesian method as exemplified by Descartes' four rules is but one restricted implementation of the universal method, presumably the one that Descartes finds compelling to obtain true, rational knowledge. Likewise, Newton's four guiding principles for reasoning in philosophy, Quine's list of heuristics that count toward the plausibility of a hypothesis, Bertrand Russell's proposed ten commandments, and Hume's three tools to guide philosophical inquiry (his microscope, his razor, and his fork) are only other specific implementations or subsets of the method to which various philosophers have left their name. Similarly, the scientific method is "use scientific heuristics"; the writer's method is "use the heuristics of the writer." Naturally, a painter uses the painter's heuristics; a mystic, those of mysticism. All methods, without exception, are fuzzy subsets of the one universal method *Use heuristics*.

---

*Lydia Rollins, my mother-in-law and a Latin teacher, was the translator of this term.

The *universale organum* is also comprehensive. In a world in which everything is heuristic, it is hard to conceive of an answer to the question "What do I do?" that is not equivalent to "Use heuristics."

Unlike the Cartesian method, the method does not depend on a prior commitment as to the specific nature of man, but emerges from a philosophical position that is sufficiently general to include all possible interpretations. No matter what heuristics are contained in the fuzzy subset used to define *man*, the heuristics in the fuzzy subset labeled *the method* are unaffected. Even to the combative and wildly hypothetical definition of man "as the organism that does not use heuristics," we can always respond, "That definition of man and the feeling that it refutes our definition of universal method are interesting heuristics," followed by the longest of pauses.

The *universale organum* is also broad enough to include the definition, interpretation, and operation of the method itself. If everything is heuristic, so is the act of defining, the definition of the universal method as the use of heuristics, and the interpretation of the words used to express this definition. All of the confusions, defects, and ambiguities of other formulations of universal method fall away when it is defined as the use of heuristics.

ENGINEERING, PHILOSOPHY, AND THE UNIVERSAL METHOD

Perhaps surprisingly, the engineering method was chosen as the model for the *universale organum* instead of the method of philosophy. This choice was made for a variety of reasons. First, the engineering method is undeniably a method. We validate it every time we drive a car, pick up the telephone, sit down at a computer, cross a bridge, take an airplane, turn on the lights, take our medicine, paint our house, put on a sweater, or fire a gun. If a universal method is to exist, certainly the engineering method is a proven exemplar.

The engineering method is a persuasive example of the universal method as it has been defined here. The engineer has been using rules of thumb, hints, best engineering judgment, and heuristics since the birth of man and has made no apology for doing so. Never has he measured his best against the shaky concepts of the truth, the absolute, or the universal. He has always been satisfied to measure it against his best view of what constitutes society's best. *No method that aspires to be universal can be; the engineering method, by not aspiring to be, is.*

Finally, establishing the proper relationship between the universal method and philosophy is difficult because the boundary of the fuzzy set called philosophy is so diffuse and has changed so much over time. If we, as did Plato, consider that philosophy has its own method (dialectic) and its own content (the search for the ultimate reality), philosophy is a subset of the method. The same can be said if philosophy is limited to intuition (Bergson), uncovering nonsense (Wittgenstein), a phenomenological description (Husserl), or any of the other proprietary definitions. Unlike the philosopher, the engineer is not given to introspection. He doesn't talk about method; he uses it. As a result, if we use the engineering method as the model for the universal method we do not first have to weed an overgrown garden to find the one true flower. The engineering method was used as a model upon which to build because it is clearly a method, consistent with our definition, and has changed little over the life of mankind.

The engineer has his hands full with the physical world and has no taste for the upbringing of his offspring, the *universale organum*. Does philosophy really want to abandon its traditional generality and accept any of the restricted definitions of philosophy given here? If not, it is well to remember that the engineering method built each of these ivory towers and, if the occupant desires, can always modify them to reveal a more encompassing view.

How many times have we seen a family that is old, set in its ways, and nearly dead regain a youthful vigor when a child arrives on the doorstep? Even if the infant is born of another, orphaned, and abandoned, its presence adds a new spark of life in its adopted home. Philosophy can revitalize itself and regain its proper generality by becoming the guardian of the method. The word *philosophy* is composed of two Greek roots: the first meaning "love of" and the second usually translated as "wisdom." In the traditional view, philosophy is the "love of wisdom." According to *The Encyclopedia of Philosophy* (ed. Paul Edwards, 1967), in the *Iliad*, Homer uses the second Greek word *sophia* to refer to the skill of a carpenter and, with only a small extension, in addition to the love of wisdom, philosophy includes the love of the heuristics of the carpenter, the craftsperson, the engineer. *Philosophy becomes the study of the heuristic by heuristics.*

This does not mean that all philosophers should abandon their traditional sotas. Each particular philosopher remains the proud parent of his own creation. Those who accept the absolute as absolute; those who believe in critical discussion and reason; those who wish to study language are free to raise their children. But for those with free time, much work remains to be done, and, if the fears of some over our mental health as a species, unbridled terrorism, genetic engineering, pollution, and overpopulation are justified, there is little time to do it.

A liberally educated person is one who is conversant and at ease with the major currents of thought of his day. In the past this clearly required knowledge of the heuristics that defined the classics, philosophy, Greek, and Latin. Today's world is far more complicated. We will not survive as a species unless we can produce a new liberal synthesis that includes the sotas of technology and engineering with the more traditional ones. Philosophy, through the general study of the heuristic, can provide this new liberal synthesis. *What we most desperately need is a New Renaissance Philosopher to engineer our world based on the search for the best heuristics for human survival.*

Once again, let us use the heuristic of the artist and reaffirm the line of this argument to reestablish the feel of this work. For the most part, our stroll has been uphill, and it has carried us rather far together. A brief review from the beginning of our journey will assure that we have not missed our final destination from being overly engaged in conversation. We climbed the mountain to reach the pinnacle where *All; heuristic* on the side of the engineer, and we must now reexamine the footholds we have used. This review will lead naturally into Part V, where the final approximation to the philosophical basis of this work will be drawn in relief. Part VI will then consider a utopian application of the universal method and conclude our time together.

You and I are participating in Nature's latest, most magnificent experiment to see whether or not the human species with its new wrinkle *intelligence* will find itself on the main trunk of evolution. You have a vested interest in the results of this experiment and so do I. If we have any doubt about its success, we should seek the most powerful method for causing the changes that will lead to our survival. To avoid elimination, intelligence must plead its most effective defense.

Long before consciousness, long before thought, our earliest ancestors were intellectually detached from their environment, but at peace with it. From a welling up of loneliness, fear, hope, or love, they did the best they could to cause the changes they desired within the resources they had at their disposal in a complex, unknown, and, at times, even frightening world. The most conspicuous physical aspects of today's environment were built upon this early foundation using what may now be called the engineering or heuristic method. Could it be that this method, if properly understood and generalized, could help to make Nature's latest experiment, human intelligence, a success?

To evaluate this proposition the following objectives were established:

1. A situation calling for the talents of the engineer was described.
2. The engineering method was developed.
3. Examples of the heuristics the engineer uses to cause change were given.
4. The engineering method was generalized to the universal method for causing any change under uncertainty.

The world of the engineer and his method for creating it are easily described. The engineer wants change; he wants the best change; he wants the best change in a complex situation; he wants the best change in a complex situation within the available resources. One accomplishes this change by relying on the past experience with similar problems that is encoded in the rules of thumb, hints, formulae, assumptions, and concrete objects around him. In a word, the engineer relies exclusively on the heuristic. Although these heuristics do not guarantee a solution, may at times contradict each other, solve unsolvable problems, and depend on context, they provide a probable, but potentially fallible, basis for decision. For convenience, a specific set of heuristics valid at a well-defined time was called the state of the art, or sota. The heuristics in an engineer's sota vary widely in complexity and interconnectivity. Some are simple rules of thumb and orders of magnitude; others are complicated instructions that tell the engineer which heuristics to use, as well as when, and how to use them. Still other heuristics tell how we should implement the engineering method, define the engineer, and judge his work. In the world of engineering, all is heuristic.

The engineer's world is not the best world in any absolute sense, but the one judged best relative to the sota used to structure it. As a result, criticizing this world on the basis of information that is now available but that was unavailable when this world was created is unreasonable. Clairvoyance is not one of the required skills of the practicing engineer. Need I add that the engineer's characteristic calmness, untroubled by mental or emotional disquiet, results from the congruence of his personal vision of the engineering world with what he takes to be the generally accepted best engineering world?

About the sixth century B.C., with the development of the heuristic *rational thought*, Western man began to structure his world in terms of one specific sota, *science*. This set contains an increasing number of heuristics, including comprehensibility, objectivity, time, space, and causality, and it is used to justify the persuasive heuristics—external reality, truth, and Nature—behind the veil. From a distance this world looks familiar, but on close inspection it does not appear congruent with our personal world and takes on the crazy air of a carnival—a *Carnival of the Rational* produced by Thales and company—complete with a sideshow of grotesquerie. This sideshow contains all manner of booths with different aged twins, perfect clocks running at different rates, cats that will die only when the lids of their boxes are opened, hairless black holes, particles that neither exist nor do not exist, time that does not flow, space with wormholes, S-matrices, and a very mysterious Bell's Theorem, to name only a few.

Most people never enter this world but stand in awe peering through the front gate listening and nodding assent to the tales told by those who have ventured inside. Occasionally we see a precocious schoolboy, popular guidebook clutched tightly in his hand, pointing out the various curiosities along the midway to his family and friends who will soon flee back to the world outside. The bulk of those who enter to stay are young. They soon forget the world beyond the fence, if they ever knew it, and take the illusory world of their own making for an absolute, true, necessary world and become captured by it. As they go farther and farther into the carnival the way becomes so narrow and treacherous that few can negotiate it safely. Most who enter cannot forgo the constant fascination and stimulation to see the hidden machinery that makes it all work, become hypnotized by what they see, and can never escape. A few, with superior intellect, are more fortunate. As they leave by the rear exit, they turn to wonder aloud at the tinsel they have seen. All the while, the carnival's Giggling Gertie sits high above the crowd, gives a majestic shrug, and laughs and laughs.

So we turn to philosophy for peace. Dogmatists single out certain heuristics, grant them the privileged status of being above dispute, and use them to analyze other heuristics. The philosophers differ only in which heuristics they grant this special status. Some choose T̲ruth; some choose B̲eing; some choose the Ex̲ternal World, L̲anguage, or K̲nowledge. To maliciously mix metaphors, they each invite a different camel's nose under the tent and then wonder as they find that they have painted themselves into a corner. The skeptic, on the other hand, professes not to take a position and so returns unscathed from what is little more than a battle of wits. Unfortunately, he fails to recognize that to carry the day he also assumes a limited sota, grants it privileged status, and then selectively rejects parts of it. His insistence, for example, that all proof must degenerate into a vicious circle, an infinite regress, or an unwarranted assumption can only be taken to admit that a small subsota containing the heuristics proof, vicious circle, infinite regress, and so on, exists, to show a lack of imagination in creating interesting additional possibilities for the fate of a proof, and to reject capriciously one selected heuristic *proof* from this subsota on this account. Or the skeptic struggles to regain the peace before intelligence by preaching the heuristic that since all positions are equally valid you should withhold judgment, in effect assuming that judgment and rational thought exist so that he can reject them. This program cannot be made convincing. Once the skeptic adds judgment to his sota, he cannot but judge. The skeptic creates a world of choice, protests too much methinks that he does not choose, chooses, and ends in despair. Not surprisingly this flawed tactic hands the skeptic only a barren, unacceptable peace—a utopia no one would want. Can we not engineer a better one?

I think so. The engineering method serves as a model for how. Peace comes with the realization that the world in which we live is an acquired taste—one we all, as artists, paint in our own chosen styles. You use your favorite colors, subject matter, and brushstrokes, and you set up your own internal standards for validating your art. I do the same. But this is also true of the scholar and the illiterate; the banker in New York City and the aborigine in Australia; the Buddhist, the Christian, and the Jew. The vision of some painters will find its expression in the isolated brushstrokes of the modern artist, of others in the mist- and void-filled paintings of the Japanese, and of still others in the portraiture of the Eternal Artist. Doing our own best, we each paint our own small part in the most magnificent of frescoes.

So dissimilar and private are the styles in which we paint that one artist can never understand the needs and responses to these needs that generate the style of another. Ignorant of these needs, one artist is not curious about the attempted solution of another and too often becomes impatient, intolerant—or worse. As a result, our fresco becomes filled with clashing colors, sharp angles, unharmonious contrasts, and unsupported diagonals.

Since one style can never be faithfully translated into another, the only hope for reconciliation is in a higher synthesis. We must hope that the intellect is capable of recognizing and accepting that one cannot judge, but judge; that one cannot assert, but assert; that one cannot choose, but choose. By creating the heuristic *heuristic* the absolute becomes less absolute; the certain, less certain; the sure, less sure. When all is heuristic, the hard, sad, cold inhuman nature of the absolute, the certain, the sure bends to the soft human nature of the heuristic. Sitting in a cave painting the shadows cast by the fire that dance on the walls, we will have no desire to turn around to see what causes them for we know that we will only find more shadows. At last the mountains of Arcadia and the mist-wrapped mountains of the lonely traveler become one as we realize that what we paint, how we paint, that we paint, and, most importantly, why we paint are all art—all; heuristic. To be alive is to be an artist; to be conscious of being an artist is to be human; to recognize and be satisfied that you can only be an artist is to be at peace.

Although this discussion is limited to only a small portion of the overall canvas, to that portion that reveals the secret of construction of the whole, it depends heavily on the art of others and requires a receptive viewer to achieve its full effect. The focal point is the kōan with which this section began and more particularly with one specific detail of this focal point, the *universale organum*. Like a painting, the construction of this discussion has not been linear. An idea, like a color, gains in vividness by close juxtaposition to another. Each dollop is incomplete in itself but each adds to the impression of the whole when brought into contact with its neighbor. Clearly, I paint an abstraction, an oversimplification, that is incomplete without the receptive observer to supply the subjective lines, mix and merge the colors, and extrapolate beyond language.

So at last, we sit high, secure, and at equilibrium on a mountain. Peace becomes peace and all departures from it only additional heuristics. Some try to sit next to us by becoming lyrical, mystical, religious, or speechless. We maintain our position by remembering that

> Before intelligence,
> mountains were mountains,
> With intelligence,
> mountains became mountains,
> With the heuristic,
> mountains became mountains, and
> After the heuristic,
> mountains became once again mountains.

We now understand clearly the philosophy of Micromégas, can find our way with the Bellman's map, and no longer need the artificial contrast to see the moon on the Zen vase. The heuristic . . .

was a boojum, you see.

The grin . . . disappears, as does
the ouroboros . . .

undifferentiated aesthetic continuum . . .

the pearl appears on a milk white brow . . .

language
decision
best

probable

provisional

all

heuristic . . .

V

Summary of The Method

VI

Application of the Method

I'm hungry. When I'm hungry, I eat. Before we continue, I think I'll go get a sandwich, and, perhaps, you would like to do the same. I'm also impatient. On all sides, prophets of doom vie to become the first modern Cassandra. When I hear them speak, I develop the engineer's characteristic sense of urgency. Can the human species really destroy itself twelve times over? I hope not, for evolution has not adapted man to be an Atlas. Who? How many? And what is the mental health of those into whose hands is thrust the power to unleash this force? I also worry about a species that has members who resort to unbridled terriorism. I become equally concerned when I stop to consider genetic engineering and the possibility of an escaped virus destroying us all. I worry when I stop to consider the self-inflicted slow death of our species as pollution and overpopulation change our hospitable ecosphere into a subtler torturer than any of the Inquisition. We must not consider these problems now, however, for they are only what concern me. I must restrict my interest to what concerns you instead. Surely there is something you would see changed. I ask you yet again: Would you establish global peace or squelch an impending family feud? Would you improve your nation or modify your local government? Would you feed the one quarter of the world population that goes to bed hungry or decide what to serve for dinner? If so, the world is not as you would have it, and you are also interested in change.

TRADITIONAL UTOPIA

This desire for change is not new. Humankind seems to have always lived with the search for a better world. Most formal attempts to describe one, beginning with a description of a Golden Age by Hesiod about 750 B.C., have been within the Western tradition. This literary genre had to await More's *Utopia* in 1516 to find its definitive name, which was then applied retroactively to these earlier accounts of a perfect society. We can hardly claim originality in wanting a different, better world and writing of our dreams for one.

We have come to expect a traditional utopia to have several characteristics: a specific location, an artifice (or frame) to transport us to it, and a description of a world that is sometimes better, sometimes worse, but in any event different from our own. At one time or another, utopia has been found on an isolated island, on a floating island, on the moon, in an area surrounded by impassable mountains, or in a distant period of time. The access to utopia has been by ship-wreck, by dreams, or by technology such as a time machine. Once in utopia, we are treated to a wide variety of customs. We see square, circular, or star-shaped cities. There are governments by rulers, committees, or complex groups. We see people who wear blue robes and green underwear; those who wear black robes and capes, helmets, and peach-colored velvet shoes; or savages who run around naked. Imagination has not been spared in the choice of location, means of ac-cess, or customs of these creations.

UTOPIA AS PROGRAM FOR CHANGE

Although these stories are fascinating reading if written by a perceptive author, traditional utopias leave much to be desired as a program for change because they violate nearly every known heuristic of engineering. All ancient utopias and most modern ones present us with a stable society in a stable environment. They have little sense of continual change, even less sense of transition from our world to theirs, no sense of history, and an obnoxious sense of self-righteous-ness. Specifically:

1. The few utopists who admit that the world evolves and hence propose utopias that are dynamic fail to include specific heuristics internal to their world to maintain, control, and evaluate these changes. That is, they give no *feedback loops* to guarantee stability.

2. The modern utopia that recognizes that we must convert the present world into the utopian one proposes large arbitrary steps instead of heuristics to promote a continuous transition. That is, these steps are nei-ther *small changes in the state of the art* nor *attacks on the weak links*.

3. In none of these visions do we sense the remnants of our world made better. This is unfortunate because sound engineering practice makes it clear that good design must build on the current *state of the art*. The au-tomobile is a direct descendent of the four-wheeled cart, and the mod-ern bridge is undoubtedly an outgrowth of a log that chanced to fall across a stream. That is, the traditional utopia does not betray a family resemblance to our present world.

4. Finally, the inhabitants of utopia plunge into their unknown world with no hesitation, no lingering doubts, and no backward glance. That is, they do not give themselves a *chance to retreat*, for error in their vision is un-thinkable and retreat, therefore, unnecessary.

No traditional utopia offers a real program for change consistent with the most powerful method for change ever invented.

At best, the utopist announces that the world is not to his liking and gives us a few sketchy, unrelated goals that he would like to see realized in the future. Unfortunately, even these goals offer little help in creating a better world, for they differ fundamentally from engineering goals in a variety of important ways. First, they are too strange, too divorced from reality, and too unreal. Is it really believable that a utopia of cannibals would have no manner of traffic; no knowl-edge of letters; no science of numbers; no name of magistrate or statesperson;

no use for slaves; neither wealth nor poverty; no contracts, succession, bourn, bound of land, tilth, or vineyard; no occupation but that of idleness; only a general respect of parents; no clothing; no agriculture; no use of metal, corn, or wine, or oil? Is it really believable that the very words denoting falsehood, treachery, dissimulation, avarice, envy, detraction, and pardon would be unheard of? The typical utopia is so foreign and represents such a large change in the state of the art that it seems curiously irrelevant and, as with this society of cannibals, is easily dismissed as little more than a tempest in a secondhand kettle.

Second, utopian goals are the eccentric whims of their authors; engineering goals are self-fulfilling prophecies. The engineer insists that a goal must be the embodiment or shorthand for a subset of the current state of the art to be useful. This subset always includes heuristics to assure its own feasibility. Because of this difference, achievement of engineering goals is nearly 100 percent; achievement of utopian ones, nearly zero.

As it turns out, the high failure rate of the latter is a good thing because the third difference in utopian and engineering goals is that those of the utopist are potentially dangerous. Consider any change you would like to see made: saving the whales, providing sufficient food for all, or whatever. I happen to have a magic wand, or if you prefer a monkey's paw, of a very special kind. If you wave it, it will instantaneously bring about the future you desire. Would you use it now to bring about your utopia?

I hope you would not, for no respectable engineer would. The world is so complex that it defies complete analysis. No one can know the implications of a small change, much less a large one. Experiments in which man has made even a small change in the ecosystem to solve a perceived problem have often produced disastrous results. In one instance, the European sparrow was introduced into America as a destroyer of insects but is now commonly regarded as a pest. Likewise, no one knows what would be the result of suddenly modifying the transient balance that Nature has established between the whaling industry and the number of whales. Since whales eat large amounts of plankton and some scientists estimate that as much as 90 percent of the oxygen in the atmosphere fixed by photosynthesis comes from the one-celled plants and animals of the ocean, a sudden reduction of the population constraints on whales could prove disastrous. Waving the magic wand to provide enough food for all of the world's population holds the same potential for tragedy. Food is a constraint on population in many parts of the world. Removing this constraint over a short period of time could cause more deaths in the long run as the Malthusian pinch simply squeezes later and at a higher population level. The same situation is true of all utopias. Would Plato, More, Rabelais, Bacon, Campanella, and Andreae, to limit ourselves to the most conspicuous examples out of the thousands of utopists, wave my wand to bring about their Republic, Utopia, Abbey of Theleme, New Atlantis, City of the Sun, and Christianopollis? If they would, fitting punishment for their rashness would be for them to be condemned to live in their creation. No engineer would choose to live in a place that was built in a day, that destroyed so many of the stabilizing feedback loops in his present world, and that provided so few new ones.

I do not see how anyone could be against saving the whales; I do not see how anyone could be against feeding the hungry; I do not see how anyone could be against many of the goals of the utopists, but that is not the point. Before I roll up my sleeves and go to work to achieve your preferred world, I insist on a set of convincing heuristics to bring about a stable transition from the present world to your more desirable one. If you want desirable change in a complex, poorly understood system within the available resources, the engineer suggests that you use appropriate heuristics for change—not a magic wand.

To summarize: The goals of the utopia fail as a program for change because they neglect to tell us how to get there from here; they have no sense of here; and they have only a pale sense of there. As a result, I, for one, am afraid to commit myself in advance to live in utopia even if we could somehow manage to bring it about. Utopias, though fun to read, are practically worthless as blueprints for change.

EUTOPIA

To replace the utopia, I propose a new instrument of change. I have read somewhere that when Thomas More named his utopia, he cleverly played on an ambiguity in the Greek language. Did he mean for us to read *utopia* as *outopia*, meaning "no place," or as *eutopia*, meaning "good place?" I can accept no doubt or ambiguity. I insist on a *eutopia*, one based on the method, on the *universale organum*.

The locus of your eutopia, your better world, the one you would bring about, is in your personal sota. That is its natural homeland; that is where it is secure and absolutely convincing. Thomas More daydreamed that he was King Utopus, the ruler of his perfect society. Francis Bacon undoubtedly created the ruler of the New Atlantis in his own image as head of a complex hierarchy of scientists. Can we doubt that B. F. Skinner saw himself as Fraiser, the master architect of Walden Two? In a similar way you are the ruler, the king, the architect of the eutopia in your own sota, absolutely safe from being deposed. I am the same.

The heuristics in your kingdom may be partitioned into three subsets. Eutopia will not be found in the first, which defines the way you presently see the world, for surely you are interested in a better one. Eutopia will not be found in the second, which defines a future, better world, because this sota will change in time and is only a pale image of the future. Eutopia is to be found in the third set, which defines the heuristics for change, for transition, for method. Eutopia is not found in your artifacts either present or future, but in your art.

Most people would be unhappy living alone in their private eutopia. They feel obligated to preach their personal vision to another and in so doing increase their landholding. Indeed, an effort to do so is often a good heuristic for change. Unfortunately, your eutopia must forever remain another's utopia, for trying to externalize your eutopia always suffers from abstraction. The words and images you use to define your perfect world are distorted, deformed, and reinterpreted by the heuristics already present in the sota of your listener. What seems so evident, so obvious in its native land seems so quaint, so curious on foreign soil. The best method for minimizing this problem is to keep the overlap between your sota and that of your listener as large as possible and to emphasize the heuristics of change, which are evaluated in the present, instead of the heuristics that define goals, which are evaluated in the inaccessible future. A eutopia intended as land development must seem familiar to the prospective buyer.

All that remains of our discussion is for me to describe my eutopia; that is, my implementation of universal method. I include it in the hope that it may contain elements that you will want to incorporate into your own. Of course, you already know its location and have probably guessed that I have intended this whole discussion as both frame and means of transport to it. To be consistent with literary practice, I need a name for my creation, one that, like the eutopia it represents, is but a small change in the current state of the art and, hence, almost recognizable. *Mundus* will serve my purpose well. It has the ring of my heritage and a certain sense of generality that pleases me well. Won't you join me for a visit to Mundus?

MUNDUS INSTITUTE OF TECHNOLOGY

The city of Mundus, one of the world's busiest ports and most important cultural centers, is reputed to be the hub of the universe, although its partisans may be the only ones to share this sentiment. I do not need to describe it for you in detail for you know it well. Just look around you now, and you will have a good idea of the language, clothing, and customs in Mundus. Men, women, and children pass us wearing many different kinds of garments, but they do not look at all strange. Unlike the sense of dislocation a tourist feels in a foreign country, here the creases in the men's suits, the hem lengths of the women's dresses, and the styles of the children's hair are pretty much as expected. In Mundus, everything seems as it should, different and varied, but also comfortable and familiar.

This is not to say that Mundus is an idyllic utopia with no problems. Its citizens struggle to establish peace, educate their children, and provide a good life for all. But what problems they now have are well on the way to a solution. The people of Mundus are happy because they emphasize desirable change over goals and are rewarded by this change as much as by the attainment of the goals themselves. In effect, they have learned the wisdom of the well-known heuristic: *The road is always better than the inn.*

What most interests me now and is the object of fascination to all visitors is over that bridge. You can just make it out on the right, behind the sailboats on the river. This large complex of buildings, the greatest jewel of Mundus, is the noblest foundation that ever was upon the earth. It is dedicated to education, but to study and learning of a special kind. The scope, generality, and importance of this institute are breathtaking and had best be approached in stages. As we complete our walk across the bridge, I propose to explain first a little of its history and how it got its name. Since its architecture is most unusual, that will occupy us next. Once inside the doors, I will consider the several employments and functions to which the people who work here are assigned, and finally we will visit some of the corridors and rooms that I think will be of special interest to you.

Origins

Legend has it that on the ancient island of Bensalem there stood a large foundation, called Salomon's House, dedicated to the knowledge of causes and secret motions of things; and enlarging of the bounds of the human empire, to the effecting of all things possible. In the old days the folks of Mundus held that their foundation, originally dedicated to the knowledge of the means used by a people to provide itself with the objects of a material culture, was derived from this ancient Salomon's House. How, they asked, could you provide for a material culture without knowledge of the causes and secret motions of things? They named their foundation after their town.

Soon they realized three things. First, technology included not only the techniques concerned with material objects, but also those that caused changes of all kinds. Second, these changes had been taking place long before Salomon's House was built, and, indeed, Salomon's House was, itself, a monument to and relied on some of the most powerful of these early techniques. Third, the word *technology* was often misunderstood. As a result, the Institute was rededicated as a place for the study of the methods for causing desirable change of all kinds. The activities of Salomon's House were relocated into one of the interior rooms. The foundation, also, became more widely known either by its initials or, more simply, as the Institute or, occasionally, as Tech.

Architecture

At first glance the Institute's architecture appears undistinguished, but it is actually quite functional. Tech consists of a number of interconnected buildings surrounding a courtyard on three sides that is appropriately called the Grand Court. Since the winters are often harsh, this design allows the many individuals who work there to pass from one area to another with a minimum of inconvenience. I believe the overall style is neoclassical, but, in any event, it is massive and conveys a sense of stability and purpose. To soften the lines, the Grand Court is planted with a large number of flowering bushes and shrubs so chosen that some are in bloom at all seasons of the year. This motif is repeated in miniature throughout the Institute in a series of pocket parks intended for repose. No expense was spared in keeping with the general rule of thumb that a person's environment is a major factor determining his happiness and the effectiveness of his effort. At times of great national crisis, spontaneous meetings have occurred in the Grand Court as students, faculty, staff, and citizens have come together in a increased sense of community. The second purpose of the Grand Court is to serve as a transition from Institute to the world beyond. In pleasant weather, the citizens of Mundus gather there to chat and discuss the current work in progress. At times this makes it impossible to tell where the dividing line between Institute and world outside really is. This confusion is heightened because no doors are visible between the Grand Court and the interior of Tech. By a triumph of modern engineering, jets of air coming from the sides of the doorframe maintain an invisible barrier between the inside air-conditioned environment and the outside one. This is but one example of purpose built into the architecture of the Institute.

The interior of Tech is no less functional than the outside, but, if anything, it is even more peculiar. Immediately inside of the Grand Court and separated from the rest of the main part of the Institute is a large antechamber. For ease in referring to this central area, I will call it the Vestibule. Corridors radiate in all directions from this Vestibule. Both sides of each corridor are lined with rooms, which may be further subdivided into subrooms, and so on. The binary division into antechamber and room proper that was found between the

Vestibule and the rest of the Institute is characteristic of all subdivisions of the building. In a similar fashion, each corridor, each room, and each subroom is partitioned into these two parts.

The styles of these corridors and rooms are all different. Some are elaborate; others, more austere. There are rooms piled high with laboratory equipment, rooms with books, and rooms with computers. Some rooms are absolutely empty; others defy description. Each room at Tech has been especially designed to achieve a specific purpose, goal, or ideal. But just as it would be difficult to study the use of computers in a room without computers, each room contains the means to accomplish its goals within it.

Above each door, including those to each room, to each corridor, and to the main entrance, is a doorplate, identical in size and shape, on which is inscribed two things: a complete list of the heuristics that are used within and the names of the principal researchers responsible for the discovery of these heuristics. For example, above the room I mentioned earlier as having been patterned after Salomon's House, the doorplate lists the heuristics *The world is comprehensible* and *There is an objective world.* Among the names it contains are those of Thales, Anaximander, and Anaximenes. Similarly, the doorplate of the outside entrance to the Institute contains what was once undoubtedly *All is heuristic,* but time has washed away the *is,* and it is hard at work on the *all.* Unless a repairman comes to its rescue, I fear that the day will come when the inscription will be obliterated for good. This motto is in keeping with the overall goal of the Institute as the study of heuristics. Somewhere, on the doorplate to some room or to some corridor, any heuristic you might care to name has its place.

These doorplates serve a useful function. Everyone who enters Tech is issued a pass listing all of the heuristics he knows. Unlike a regular permit, whose principal function is to restrict entrance, this pass serves to remind its owner how he got into a specific area. As a visitor penetrates corridor, then room, and finally subroom, he inserts his pass into a slot in the doorplate of each and the specific subset of heuristics he needs to enter is automatically registered on it. Leaving a room causes the reverse process to take place. This system of doorplates was developed when Tech became so complicated that it was nearly impossible to remember how you got into a specific room.

One of the oddest things about the Institute is that I cannot tell you how many rooms and corridors it contains. My embarrassment is for four reasons. First and simplest, rooms are constantly being added to and removed from Tech. In other words, the number of study areas is constantly changing. Second, and more seriously, the walls and doors are not distinct, but indeterminate and fuzzy. A legitimate dispute can arise as to what constitutes the boundaries of a room. One person may hold that neighboring areas are sufficiently similar to be included within the same room, and another may insist that two separate rooms are required. Unfortunately, no higher authority than individual opinion exists by which this disagreement can be adjudicated. The third reason for my difficulty is that a room can be entered through different doors making counting difficult. A complex network of tunnels, secret passages, and doors interconnects the corridors, rooms, and subrooms. Like the rooms themselves, these entrances are constantly changing. To try to establish a definitive count of the number of rooms in Tech under these fluid conditions is a hopeless task. In principle, a person can enter any doorway he sees, but therein is the most peculiar feature of the Institute and the fourth and final reason why I cannot tell you how many rooms the Institute contains. While it is true that a person can enter any door he sees, it is not true that he can see all doors. Only those areas that correspond to the heuristics on his pass are visible. He may enter rooms in which he does not know what is going on, but only on the condition that at least he knows that he does not know what is going on. Obviously, this causes a final insurmountable problem for anyone who wants to determine the correct number of rooms in the building.

Personnel

I see we have finished our trip across the bridge. By now you are familiar with the general layout of the building, and I am eager to consider the people who work here and to peek in on a few selected rooms.

Do you see that small group of people huddled to the right of the front door? In former times this group was much larger, but of late their number has dwindled. They seek one thing that is not a heuristic, and for this reason they must sit outside of Tech. You might think that they represent the dogmatic philosophers of old, but this is not the case. Dogmatism is but yet one more heuristic and occupies a place of honor along with skepticism inside the Institute. This small group is concerned with the basic premise of the Institute itself. Their search is not an easy one, for as I said the weather in Mundus is cold in the winter. But they persist despite sadness in their quest.

Now we pass through the front door and into the central Vestibule. Let's sit down and rest awhile, and I'll explain the responsibilities of the people who work at Tech.

Among the people who are needed to accomplish the goals of the Institute, three different kinds of people are easily identified: the Abstractors, the Professors, and the Students. As with the rooms and corridors, the boundaries between these categories are fuzzy, but each of them has specific functions that should be of interest to us.

Abstractors

First let us consider the group called Abstractors. They are assigned to the antechambers of each corridor, room, and subroom and have similar responsibilities. They accumulate the heuristics provided from more specific levels; analyze, study, and generalize across the spectrum of heuristics they receive; and pass their conclusions on to the next higher levels. The conclusions reached by the Abstractors who work in this central Vestibule are given, as expected, to the population of Mundus. The communication between Abstractors is on a monthly basis with the major flow of information from the more specialized rooms to this central area. Since the Vestibule is of special interest, I'll describe the work that goes on there in more detail in a minute.

The importance of the Abstractors was recognized early. In the old House of Salomon there were groups of men charged with collecting the experiments given in all books. For reasons that I do not understand, they were called Depredators. Other men called Compilers were responsible for arranging the experiments of others into tables in order to aid in drawing observations and axioms from them. Later, large companies offered abstracting services and published their results in books such as the *Chemical, Social Science,* and *Nuclear Ab-*

stracts. The objective of these books was to report in a shortened but still technical form the content of all articles published in a given area. Typically, only a professional chemist could read, much less understand, the *Chemical Abstracts.*

The work of the Abstractors at Tech serves somewhat the same function as that of the Depredators, Compilers, and former abstracting services, but it is more general. The Abstractors seek, for example, a heuristic analysis of as many different situations and as wide a distribution of their findings as possible. A heuristic analysis includes a compilation of the set of heuristics that represents the current state of the art as expected, but it also includes a list of those heuristics that are leaving the state of the art as no longer useful, those entering it, and the heuristics that are promoting these changes. In other words, a heuristic analysis preserves the dynamic character of the state of the art and explicitly identifies the forces that cause it. How many paintings, how many musical compositions, how many technical advances, and how many governmental systems have never been discovered because the heuristics known to the individual who could have made these advances were insufficient? Every flower that fails to bloom because ignorance sowed it in the wrong soil is a tragedy for the human species. The function of the Abstractors is not only to collect information in a technical form to enable researchers to retrieve it quickly, but, even more importantly, to make it widely available in a variety of forms to minimize this tragedy. The principal task of all Abstractors at Tech is to make the state of the art of all areas accessible to the widest possible number of people.

Providing information to the Abstractors is the task of everyone. What is needed is a heuristic analysis of all areas of experience, and only a person knowledgeable in each area can draw up a relevant list of heuristics and constantly update it. Even now we desperately need a first cut at the set of heuristics for establishing peace, electing worthy leaders, and ensuring human survival. We do not as yet even have the best state of the art for law, politics, medicine, education—not to mention—child raising, city planning, and human relations. The Abstractors seek not a definitive list, but the best thoughts of anyone who has information to contribute. We are all responsible for providing the state of the art in our own area to the Abstractors.

Professors

As in olden times, the people who work in the principal part of each room are called Professors. They form the second recognizable group at Tech. Their responsibility, as might also be expected, is to research and to teach the best heuristics our species has discovered to the next generation. What is less expected is the breadth of knowledge the Professor is responsible for having.

Few Professors are allowed to become experts in one isolated area, and those who insist on doing so are not among the most highly respected members of the intellectual community. Each Professor is encouraged to have at least two (preferably three) subjects in which he is genuinely qualified as an expert—not at the level of the Abstractor, but at the level of the Professor. In the past, dividing knowledge into separate areas and encouraging specialization in one of them was considered a good heuristic. A chemical engineer knew what journals to read, what conferences to attend, and who to call on for additional information in chemical engineering. In this way he could easily remain at the state of the art in his field. It developed, however, that excessive specialization was lead-ing to the problem of the captured expert, and a new trade-off between depth and breadth had to be found. Nowadays the hyphenated Professor holds sway. The engineering wing has philosopher-engineers, musician-engineers, educator-engineers, artist-engineers, and ecologist-engineers, and the philosophy wing has engineering-philosophers, scientific-philosophers, and so forth. Each corridor acts as if it alone must carry the total of human knowledge on its shoulders.

Encouraging Professors to become knowledgeable in and to contribute to more than one area was easily achieved. Numerous dual, endowed chairs and joint appointments were established between the various corridors, and the Professors who held one of them received a significant increase in salary and in respect. The title of Institute Professor was created and reserved for the holder of one of these chairs. It is the highest Professorial rank at Tech and understandably much sought after.

Students

The same demand for breadth extends to the last identifiable group at Tech, the Students. Recognition of the importance of broadly educated Students is not, of course, new. What is new are the strategies Tech uses to attain this goal. In the past, additional courses were added to the required curriculum to force Students to study in a variety of areas. In some cases the accreditation of a degree depended on the proper ratio between courses. This proved ineffectual because every course added either removed an important course from the major area of study or increased the time needed to obtain a degree. In addition, the best Professors were seldom assigned to teach the courses outside of the major, and advisors always suggested the easiest possible nontechnical courses so as not to affect the Student's grade point average. I know of one case in which an engineering Student was not even allowed to take a course in French because it was not closely enough related to his major. What should have been an obvious rule of thumb from the start was discovered at Tech. The most effective way to get wise, scholarly, well-read Students is to have wise, scholarly, well-read Professors. Now with the breadth required of all Professors, the Student in, say, engineering naturally comes into contact with the philosopher-engineer, the musician-engineer, and so forth, and automatically expands his knowledge in these areas.

A second strategy is also used to achieve scholarly Students. The new Student to the Institute is responsible for visiting each one of the major antechambers both before choosing his field of study and periodically during his college career. This is not treated as a routine, token exercise because one of the requirements of the Student who ultimately receives the highest degree the Institute has to offer is that he be conversant with all of the major areas of knowledge at the level of the Abstractors. No longer will the graduate nuclear engineer be considered an educated person, certified to decide questions concerning nuclear energy generation on behalf of society, or endorsed to design and build nuclear reactors if he is not familiar with ecology, value philosophy, and human psychology. No longer will the graduate liberal arts student be considered liberally educated, certified to select our artistic and literary heritage, and endorsed as a guardian of our humanity if he cannot explain what an engineer means by a trade-off, an optimum, and the state of the art. The prestige of the Institute, carried in its highest degree, will not be granted to any Student for whom Freud,

Carnot, Descartes, Skinner, Watt, Yang, or Kant is but a name on the spine of a book.

The overall objective of the Institute is to establish a sense of community dedicated to the discovery, improvement, and teaching of the heuristics that determine our preferred world. Since the boundaries that define each area are fuzzy, since the heuristics in each area change in time, and since the heuristics in each area are interrelated to others outside its own domain, this goal cannot be achieved by isolated individuals working as experts in ever-narrowing fields. By attracting wise Professors with a broad view to teach, wise Abstractors to accumulate and disseminate knowledge, and wise Students to learn; by encouraging those who were already present at the Institute to broaden their view; and by dissuading those who insist on remaining narrow from staying, the goal of a community of scholars with overlapping fields of interest has for the most part been achieved.

RESEARCH IN PROGRESS

When we first sat down, I promised to discuss some of the research that was in progress in the Vestibule. Now is the time to make good on that promise. Then we will take the opportunity to visit some of the most interesting corridors that radiate from this central area.

Research in the Vestibule

The principal activity in the main Vestibule is the study of the heuristic. Because the research in this area is so extensive, I will only have time to mention five of the most important topics currently under investigation. The first concerns the problem of making people recognize the universal nature of the heuristic. This work is well along at present, as evidenced by the existence of the Institute, and so I will pass directly to the second topic. Once a heuristic analysis is available for a wide variety of human situations, a comparison of the different lists reveals rules of thumb of general applicability. This task is actually a generalization of the work of the Compilers in ancient Bensalem. For example, the heuristic (first discovered in the corridor that deals with evolution) *Diversity and adaptability increase the probability that a species will survive in a changing environment* has applications outside of biology. A student who overspecializes in school decreases his possibility of finding employment upon graduation. An investment portfolio with a diversified list of stocks has maximum protection against a changing business climate. A company decreases its vulnerability by making the work units more interchangeable. All of these examples show that evolution's rule of thumb concerning diversity and adaptability is shared by a wide variety of systems. Once heuristics are available from many different areas of study, the Abstractors can begin to compare, to evaluate, and to generalize.

A third topic of study begins with a question. Can heuristics be allowed to evolve randomly, as they have done in the past, or must we develop heuristics to control the evolution of heuristics? This problem was first recognized in the engineering wing, but later became a more general concern. Since its importance is not well appreciated, we will do well to consider the engineering experience in some detail.

The past engineer caused changes that were significantly different from those of today in five important ways: Past changes were more independent, smaller in magnitude, more limited in space, more localized in time, and more standard. Not so long ago, the engineer designed isolated, relatively small projects. He built a dam here, a bridge there, and a chemical plant some distance away. The influence of these projects on each other and on the biosphere was strictly limited. A stream really did renew itself every ten miles, and the oceans were effectively infinite because the engineer never came close to exceeding the buffering capacity of the ecosystem. When only ten Ford automobiles existed in the United States, assuming that the American automobile did not affect the life in China was a good engineering heuristic. Not only did a specific automobile design perturb a limited space, but also it did not persist over time. Although it may seem that buildings of the past lasted longer than those of today, on the scale of human evolution, both are temporary. Even the nose of the Great Sphinx at Al Jizah hasn't made it into the present day. Engineering projects of the past were also more uniform and reproducible. All ten of those Fords were black, and most of the knowledge gained in the design of a bridge in Maine was directly applicable to the construction of a bridge in Texas. Minor errors found in one design could be corrected in the next, and bridge design evolved smoothly. The heuristics used by past design engineers were based on these givens.

Things have now changed. Current engineering projects are more interdependent and tightly coupled, and they make larger incursions in the biosphere. No longer one chemical plant, but hundreds line the banks of the streams. The rejected heat and waste from one affects the performance of its neighbors downstream. In a similar way, one building in a large city with special reflecting glass can completely change the air conditioning requirements of its neighbors. If an engineering decision has an adverse effect, it is no longer local but often global. The pesticide DDT (p,p'-dichlorodiphenyl-trichloroethane) sprayed on a farm in Idaho has appeared in the bones of penguins in Antarctica, and small gold needles dispersed in the stratosphere by one country to develop a communication network could potentially disrupt communication in another. The ultimate example of this change of scale, I suppose, is the atomic bomb. Manufactured in one location, it is capable of completely devastating a spot thousands of miles away. Engineering products are also more persistent in time than they were in the past. Radioactivity introduced into the biosphere and modifications in the ozone layer as a result of engineering activity will be around much longer than the nose of the sphinx. In many cases, the changes caused by the engineer are not standard, reproducible designs that have evolved and been tested over generations, but one of a kind, irreversible, and definitive. Figuratively speaking, can one red button somewhere really unleash twelve times the destructive power needed to obliterate the human species and terminate Nature's experiment with intelligent life once and for all? The Institute recognized these five major scale changes in engineering practice and established a small group of researchers to consider if current engineering heuristics were adequate to deal with them. If the answer was *no* (as turned out to be the case), they were further charged to develop heuristics to accommodate these changes. In effect, they sought heuristics to promote the orderly evolution of engineering heuristics.

What was once thought to be only a problem for engineering is now recognized as a more general one. Since this group was established, similar major changes in the relationships between nations, large international companies, races, and religions as a result of improved communication networks, food shortages, more powerful weaponry, and so forth have been identified. An active project is underway to develop heuristics to control the evolution of the heuristics in these areas as well, and the initial group has had to be greatly enlarged.

Topic four under investigation by the Abstractors who are charged with studying the heuristic is not difficult to understand. A team of special Abstractors is hard at work developing heuristics for passing those heuristics that one generation finds useful to the next generation. Education, books, and apprenticeships are traditional heuristics in this area, but they are often ineffectual. This team seeks the most powerful heuristics to help an individual learn to learn what he needs to know.

The last project in the central Vestibule that I am aware of because of the restricted nature of my pass will not detain us long, for its importance is self-evident. A special group of researchers is charged with discovering the best way to inform the public of the work of the Institute. Since the public has an enormous influence through the courts, legislature, and ballot box on the choice, use, and effectiveness of the heuristics developed within the Institute, this responsibility is not taken lightly. If the correct basis for judging engineering design, for example, is not used, technical progress will suffer. The engineer unreasonably held responsible for the failure of a product that represents the state of the art would respond by decreasing the risk he would accept in the future. In essence the heuristics the public uses to evaluate engineering progress controls the risk-taking attitude of the engineering profession. Often this is as it should be, but if based on a lack of understanding of the art of engineering, it is not. The ultimate result is a change in the implementation of the engineering method, and as this method affects intimately how we live, the public (especially judges and juries) must be well enough informed to apply the correct rule of judgment instead of a faulty one. A technically illiterate population can cripple technical progress or stop it altogether. Because technology of all kinds is too essential to our well-being to leave in the hands of the uninformed, the Abstractors have an active program to inform the public of the major heuristics developed at the Institute.

Needless to say, the five topics I have just mentioned are not the only important ones being investigated by the Abstractors in the Vestibule, but they do give some idea of the current work in progress.

Research in Corridors

That's about all I know of the architecture of Tech, its personnel, and the activities that take place in the central Vestibule. I'm rested now. Do you think it might be interesting to visit a couple of the corridors that radiate from this Vestibule?

Let's begin with that important-looking one on the left, the one with the doorplate marked Thales. In this corridor's antechamber the Abstractors study the heuristics that define the scientific method. When compared to the methods of art, philosophy, and especially engineering, the work they do is very well advanced. Books by Aristotle, Bacon, Descartes, Kuhn, and Popper, to mention only a few, are available to aid the novice who might want to enter here. I have often mentioned the large room that you see on the right immediately upon entering this corridor. It is the one that evolved from what was once called Salomon's House, although it is now hard to recognize because it has been completely renovated. The Abstractors in its antechamber are principally concerned with the method of induction. Originally, their studies seemed to show that this strategy was powerful enough to uncover all of the secrets of Nature. As a result, they used to recommend that "Nature be put on the rack and forced to divulge his secrets." But a scholar who tunneled in from the neighboring room dedicated to philosophy destroyed the absolute power of induction (and incidentally, causality) once and for all. Since then, the Abstractors in Salomon's House have become more subdued. Caves that were dug deeply into the foundations of this room to be used for coagulations, indurations, refrigerations, and conservation of bodies still exist, but these processes have been considerably updated. For the most part, the chambers originally intended for such things as generating frogs, flies, and diverse other creatures spontaneously from air; making diverse plants rise by mixtures of earths without seeds; and changing one tree or plant into another have fallen into disuse and are now preserved as one of the curiosities of science in the room down the way that is dedicated to the history of science.

To the left just across the hall from Salomon's room is the room dedicated to physics or the science of the natural world. I have never been into the many smaller interior rooms, but my friends tell me that the things it contains are so strange that you feel as disoriented as you do at a carnival. Other rooms along this corridor are labeled chemistry, biology, and so forth. My pass is valid for only some of these, but you may be able to see and visit many others. Rather than continue any further in this direction, let's return to the main Vestibule and try a different corridor.

Obviously we do not have the time to visit all areas of the Institute. In fact, I would not even want to go into some of them, and I doubt you would either. In some the most heinous things are studied: torture, murder, genocide, infanticide, and assassination. What good could ever come from all of this, I'll never know, but the Institute prides itself that all heuristics have their niche somewhere.

Over on the left is a more interesting corridor dedicated to philosophy. Groups of men in the small inner rooms explore the great systems of the past. Of course, they consider the work of Hume, Kant, and Descartes, but they also study the great Chinese and Indian philosophers. We need not linger here too long because the work of these great system builders of the past is vaguely familiar to everyone. As before, the Abstractors in the antechambers are hard at work studying the methods of philosophy, with special emphasis on its heuristic nature. Rumor has it that the entire philosophy section will be moved to the central Vestibule since the love of heuristics, all heuristics, is common to everyone who works at Tech. No definitive decision has been made, however, and whether or not this will occur remains to be seen.

You may be surprised by the research that takes place in the next corridor. Here all of the religions of the world are being investigated from a heuristic point of view. Included are investigations of Christianity, Islam, Buddhism, Judaism, and Shintoism as expected, but consideration is also given to Paganism, Agnosticism, and Atheism. No activity that pertains to the wide variety of religious experience escapes notice.

I do not claim to serve as a very competent guide in this area. On the authority of one whom I obviously respect highly, I have been warned that to succeed in the examination of this special corridor, I would have to be more than a man and to have extraordinary assistance from heaven. Still, we would be remiss to pass by without a cautious glance into the antechamber. Perhaps in the future the Abstractors who work here will be able to make the work in this important corridor more intelligible.

From my limited understanding of the present state of the art, no conflict in the religious and heuristic viewpoints exists—quite the contrary. The philosophical position that seems to serve as the basis of the Institute suggests that everything is seen from within a limited human perspective and structured out of those heuristics that are available. Defining God, religion, and faith as heuristics does not necessarily detract from their value or existence in a world that insists that *necessarily, detract, value,* and *existence* are also human heuristics. Admission of human limitation would seem to leave unaffected the fundamental nature of a God, to enhance His uniqueness, to explain His inaccessibility, and to demonstrate both the need and existence of religion if a person's state of the art so dictates. A God is almost universally considered as a good heuristic. Your god, for others, is usually not.

Religion would then appear to be man's response to the claustrophobia induced by recognition of the constraints imposed by the set of heuristics that defines his world. Remembering these constraints reconciles many of the perplexities inherent in religion. For example, a problem certainly seems to exist if two religions each claim to be unique and at the same time the only true religion. In a heuristic world, we have simply painted ourselves into a corner by forgetting the heuristic nature of *two, unique,* and *only.* Likewise, limiting God to human heuristics when He goes about the important task of saving souls is a bit impudent, I would think. When one of the faithful is confronted with the two conflicting conditions: "You will not be saved unless you do such and such" and "A person who does not do such and such," he immediately concludes that this person will not be saved. What right does the believer have to constrain God to the use of heuristics that come from the logic system (in this case, Aristotelian logic) prominent in Western man's present state of the art? Religion soothes us and makes tolerable a world in which *All is heuristic.*

Faith is a heuristic to describe a state of the art that accepts its limitations, and believes. No person is certain, apart from faith, whether he is awake or sleeps; no person is certain, apart from faith, that nature is uniform and intelligible; no person is certain, apart from faith, that natural phenomena conform to laws of nature; no person, apart from faith, is certain that the world is under the direction of a wise and benevolent intelligence; no person is certain, apart from unquestioning faith, whether man was created by a good God, or by a wicked demon, or by chance. As Kant has said, "[Faith is] holding a concept to be true on grounds that are adequate for action, although they may not be sufficient to satisfy the intellect." That makes faith sound very much like a heuristic, a good and very necessary one at that. God, religion, faith—I know of no way to exclude them from being heuristics, nor do I find it desirable to do so.

Tech contains many other corridors, such as those in which art, literature, and language are studied. Once again I would like to pass them by, for the heuristics used in these areas are more or less well known. A more profitable use of our time will be to explore the corridor to the left. It deals with the survival of the human species, of human society, and of other human institutions.

Before this corridor was constructed at Tech, no degrees, no endowed chairs, and no books on human survival existed. What magazine articles, speeches, and studies in futurology did exist were dispersed and often based on questionable premises. Under these conditions, intelligence was on the verge of committing suicide. If Nature's experiment with his new wrinkle, the human species, were to be successful, it would have been more a matter of luck than intelligence entering into battle on its own behalf. Now all this is changed. The researchers in a special corridor study human survival, making a special effort to establish the state of the art of present society, to locate fatal flaws in the organization of human affairs, and to develop strategies to neutralize any flaws that are found. Let's look in on them briefly to get an idea of their work.

Many candidates have been proposed for the unit of evolution. At one time or the other the breeding individual, the taxon, a portion of the gene pool, DNA-in-the cell, and the flexible organism-in-its environment have been advocated as the basic element that survives or is selected against in the process of evolution. Each of these concepts is a heuristic depending for its definition on other heuristics that have been taken from the Western scientific tradition.

Instead of any one of these definitions, the biologists at Tech propose the general notion of a state of the art (for which they have coined the acronym *sota*) as the basic unit of evolution. This choice is sufficiently general to include all of the previous definitions as subsets. Each person is free to choose the specific subset as unit of evolution that best serves his needs. Those interested in the survival of life are concerned with continuity in sotas that contain the heuristics that satisfy their definition of life, and those interested in the survival of human life are concerned with an even more restrictive subset that also includes a satisfactory definition of human. By its generality the sota as unit of evolution accommodates all possible choices.

Individuals working together effectively increase the number of heuristics available to each. The overlap and interaction of different individual's sotas carves from the set of all heuristics a new sota that includes heuristics governing the articulation of the individual sotas. This new sota, called society, consists of complex, interrelated feedback loops. Two basic questions for humanity are whether the sota of the individual and that of society considered as units of evolution have survival value. The existence of the heuristic *intelligence* reduces the answer to one of choosing the best implementation of universal method.

Professors at Tech are concerned that either the sota of the individual, the sota of society, or the sota that defines the interaction between them may contain a fatal flaw that intelligence cannot neutralize, a fatal heuristic that will ultimately make humans an endangered species. Abstractors and Professors have advanced projects in this area. A discussion of two of these projects-in-progress, the advisability of intervention and the potential mismatch between the individual's goals and those of society, will demonstrate the need for the special study of human survival to seek such a flaw.

Whether man should intervene in the unfolding of Nature is an important consideration. Just because he knows how to modify his world doesn't mean he knows how to modify it safely. Two examples will illustrate this point: the work of the behavioral engineer and that of the genetic engineer.

The behavioral engineer is a modern behavioral psychologist who suggests the regulation of human affairs by direct intervention in establishing, managing, and scheduling the rewards a person receives. He believes that behavior can be controlled and improved in this way. The results of the behavioral engineer in modifying the behavior of animals and in treating the mentally ill are impressive. But the fundamental question is whether the present sota of behavioral heuristics is sufficiently developed to contemplate intervention on a larger scale. What assurance can be given that the right group will master the behavioral modification techniques first? History offers ample evidence that ability to control is not necessarily correlated with goodness of purpose. What assurance can be given that the presently known subset of techniques for behavioral control is sufficient to guard against harm? This science is so young that additional heuristics, not necessarily found within psychology, may be needed for safety.

To a large extent society is protected against error in this field because a large intervention is probably not feasible at this time. The same cannot be said for genetic engineering. The genetic engineer seeks to improve the various species by directly intervening in the hereditary material. If a gene is not as desired, change it. If a fetus has a gene for an incurable disease, remove it. If a species of insect destroys crops, modify the insect. Unfortunately, in this case a small intervention holds the possibility of enormous harm. The accidental creation and escape of a lethal virus from one laboratory by one man could spell the death of the human race.

Both behavioral and genetic engineers recognize that they want change in a highly complex, unknown system and, not surprisingly, instinctively appropriate the title engineer. Saying you are an engineer, however, doesn't necessarily mean that you are a very good one. The present state of the art of both the behavioral and genetic engineer contains the appropriate heuristics for behavioral modification or genetics, but few of the heuristics of engineering. The behavioral engineer knows how to schedule the reinforcement an organism receives, and the genetic engineer knows how to change inheritance, but neither has studied even the rudimentary heuristics of engineering. That is, neither has the slightest notion of the importance of making *small changes in the sota, attacking the weak link,* or *allowing a chance to retreat.* Given this situation, the Professors at Tech want to know if it is wise to allow them to intervene in our destiny.

A second project-in process investigates the conflict between the individual's goals and society's goals. Throughout history this potential flaw in human society has appeared in many forms, the most famous of which is surely Garrett Hardin's in an article in *Science* entitled "Tragedy of the Commons." Years ago in New England cows were grazed on land held in common by a village. In the city of Mundus a central area is still called "the commons." The monetary return to an individual for adding a cow to the commons, raising it to maturity, and selling it was large. On the other hand, if everyone added cows to the commons, it would become overgrazed and be unable to support any animals at all. The individual is presented with a problem. His personal good is served by adding additional cows; society's good is served if he exercises restraint. Several examples demonstrate that similar conflict occurs frequently in life. Companies profit by dumping their wastes into streams that are the common property of all, but if a large number of companies do so the ecosystem will break down. A couple that has two children may desire a third, but if everyone exceeds the replacement level of the population, a large increase in world population will result. The student who cuts across the grass profits by getting to class more quickly, but dirt paths across many campuses prove that the commons has already collapsed. Each of these examples proves the power of the tragedy of the commons.

Many solutions to this problem have been suggested: administrative law to regulate the commons, an appeal to individual's conscience, and so forth. All fail. Hardin's own solution, mutual coercion–mutually agreed upon, also fails because of a paradox discovered by Nobel laureate Kenneth Arrow. Arrow's Paradox proves that a pure democracy cannot work. In essence, we cannot agree by vote on how to coerce people to respect the commons because voting itself is flawed. In fact, no solution to the tragedy of the commons has yet been found. Some recent work seems to hold out hope that a solution to the tragedy of the commons may exist based on the very human need for fairness, but no one really knows. Is man fundamentally flawed and doomed to seek his personal goal at the expense of society's good? The Professors at the Mundus Institute of Technology badly want to know.

Of course this is not the only potential fatal flaw of man in society. Work is under way to see if . . .

THE REST HAS NOT BEEN PERFECTED

AN ANACHRONISTIC PREFACE

My students frequently ask why I bother to take time away from my engineering research to teach the course that is the basis for our present discussion. You may well ask the same question about this book. Nietzsche claims, Camus repeats, and I agree that for a philosopher (and, I would add, for an engineer) to deserve our respect, he must preach by example. The world is not as I would like—not as good, not as peaceful, and not as happy as I would hope it could be. One of the realities of the present situation is the necessity of producing our world with the knowledge and resources we now have. The engineer does not fear ignorance in action; he fears ignorance trying to justify inaction when inaction is known to be fatal. Since the world is not as I like, as is true of all engineers, I feel compelled to work to solve the problem.

This book and my classes are an attack on what I take to be the weak link, an attack where the allocation of my personal resources would seem to me to have the greatest effect. If I can convince you that everything is a heuristic, that you are constrained to use them, and that the ones you choose create the world you have, this small positive change has the potential for growing until we have a world we can share. Then at last, I achieve the peace of the engineer who knows that he has done the best he could, given the resources at his disposal, to create the world he would have.

Everyone wants to know how the magician works his magic, but everyone is disappointed when the trick is revealed. Let into the secret, some will claim to have known how the trick was done all along, others will register disappointment at its simplicity, and still others will claim that they never really wanted to know what it was in the first place. Mindful of this problem, I will risk a few words on how this book was written.

The mechanism of most books is intended to be transparent to the reader. As Nathaniel Hawthorne has instructed us, the greatest merit of style is, of course, to make the words disappear absolutely into the thought. Exceptions do exist. The theme of James Joyce's *Finnegans Wake* is cycles and so, interestingly, the book begins with the end of a sentence and ends with the beginning of this same sentence. Don Marquis' *archy and mehitabel* was supposedly written by a cockroach and contains no capital letters because a cockroach could not have reached the shift key on a typewriter. Still, by and large, the physical book and how it was written are seldom the point. In the present case, the construction of this book is itself the issue.

My intent was to weave the heuristics that define this book into the heuristics that define your world, while minimizing the irritation of those who do not share your sota. Indeed, the notion that the sota that defines each of us is unique was a central theme of this discussion. To forcefully make this point, my challenge was to write a book that would read differently for each person who came to it. I wanted the poet, the philosopher, and the physicist to read a different book. Only the poet may have heard an echo of the "Thanatopsis" lost in its pages. Only the philosopher may have heard Descartes speak in the quiet reference to clear and distinct ideas when the engineering method was first defined. Only the physicist may have appreciated a subtle, well-hidden pun from his domain. I wanted even the skeptic to read a different book from the one read by the general philosopher. The skeptic is undoubtedly more attuned to the oblique near quotation of the celebrated skeptic George Berkeley in the Introduction that was to become the first hint of things to come. In this book, the dramatist, the novelist, even the chess player has had his day. If you think you sense a unique part of your world in a word, phrase, allusion, style, or cadence, you are probably right. Although I have scrupulously tried not to borrow too freely from the thoughts of others, I admitted in the Introduction that this borrowed scenery was an essential part of this book's design. Each person will necessarily read a different book.

If my heuristics have been good ones, your present sota is different from your initial one when you first began to read. If this is so, you should consider how this change came about and why the new one is now so compelling. This book is intended as a concrete demonstration of different, compelling sotas and how they change—with you, who now hold it in your hands, as the subject. Our discussion was not intended exclusively as theory; it was intended as an application of the universal method.

This book was written to be reread. Let me repay the debt I incurred when I quoted George Berkeley in the Introduction by continuing his lengthy quotation, this time, with attribution from *Three Dialogues Between Hylas and Philonous* (The Preface).

> A treatise of this nature would require to be once read over coherently, in order to comprehend its design, the proofs, solution of difficulties, and the connexion and dispositions of its parts. If it be thought to deserve a second reading, this, I imagine, will make the entire scheme very plain . . .

Do you doubt that you would read a completely different book based on the new sota you now have if you were to begin at page one anew?

When this section of the book was first shown to one of its editors, he remarked, "Why are you telling the reader why and how the book was written here at the end? The Preface comes at the beginning." This anachronistic preface is intended as a true preface for your second reading if you choose to do so. On your second and subsequent readings you should feel free to ignore the faux Preface at the beginning of this book. It was included only to conform to the contemporary heuristics of book publishing by giving you something to read as you stood in the aisle of a bookstore.

To what end has all of our work been directed? You and I have created a method for action. If the world is not as you would have it, use heuristics, the best ones you know. If you do not have an effective set of your own, use those of the engineer augmented, if you desire, by those of rational thought, science, the liberal arts, religion, and philosophy as a first cut. Although these heuristics evolved in the course of history primarily to deal with the physical world, they are the most incisive rules of thumb known for causing desirable change in any complex, unknown system within the available resources. Many of them can be taken over intact or slightly modified to cause the change you desire. Even Nature seems compelled to have called upon most of them. Your world—what it was, what it now is, and what and if it will be—begins, depends, and ends with your heuristics. This book, like it began, ends as it must with you.

Time stamp of original idea: Boston, Massachusetts, Spring 1965
Time stamp of first publication: Gulf-Southwest section of ASEE, Ruston, Louisiana, 1971.
Time stamp of first complete draft: Austin, Texas, August 1982
Time stamp of present version: Austin, Texas, January 2003

HEURISTICS